MORE
OF

THE BEST OF THE JOURNAL OF IRREPRODUCIBLE RESULTS

SEX AS A HEAP OF
MALFUNCTIONING
RUBBLE (and further improbabilities)

EDITED BY MARC ABRAHAMS

Workman Publishing, New York

Copyright © 1993 by Blackwell Scientific Publications, Inc.
Illustrations copyright © 1993: Jack Tom (pages 20, 28, 34, 35, 43, 48, 50, 58, 59, 76, 78, 83-85, 102, 112, 152; Fred Winkowski (pages 8, 9, 24); Maurice Kessler (pages 6, 120, 160); and Catherine Lazure (pages 51, 72)

Library of Congress Cataloging-in-Publication Data

Sex as a heap of malfunctioning rubble : more of the best of the Journal of irreproducible results / edited by Mark Abrahams.
        p.   cm.
    ISBN 1-56305-312-8 (pbk.)
    1. Science—Humor.   2. Science—Miscellanea.   I. Abrahams, Marc.
II. Journal of irreproducible results.
Q167.S49   1992   92-50286
502.07—dc20   CIP

Cover illustration by Jack Tom

Workman Publishing Company, Inc.
708 Broadway
New York, NY 10003

Manufactured in the United States of America

First printing August 1993
10 9 8 7 6 5 4 3 2 1

Please Note
Most of the articles compiled in **Sex As a Heap of Malfunctioning Rubble** have appeared in the *Journal of Irreproducible Results* between the years 1989 and 1993. Some authors, when originally submitting their articles to the *Journal,* requested that fictitious names be used instead of real names. In **Sex As a Heap,** we have maintained these requests. Because this book is a compilation of articles that have appeared in the *Journal* throughout a period of years, many of the people who have written for it have changed their job titles since writing their articles. Therefore, a title appearing with an author's name may no longer reflect an author's position or whereabouts.

THE JOURNAL OF IRREPRODUCIBLE RESULTS
Official Journal of the Society for Basic Irreproducible Results

Editor
Marc Abrahams
Wisdom Simulators, Inc.

Publisher
James A. Krosschell

**Founder and
Editor Emeritus**
Dr. Alexander Kohn

Publisher Emeritus
Dr. George H. Scherr

Associate Editors
Mark Dionne
Interleaf, Inc.

Stanley J. Eigen
Northeastern University
EIGEN @ NORTHEASTERN.EDU

Michèle M. Meagher
Wisdom Simulators, Inc.

Bertha Vanatian
MIT

Assistant Editor
Stephen Drew

Contributing Editors
Karen Hopkin
Steve Nadis
Alice Shirell Kaswell

Art/Photo Contributing Editors
Lois Malone/Rich and Famous
Graphics

Roland Sharrillo

Dr. Robert Richard Smith/Minimum
Wage Art

Production Editor
Joe Sweeney

Design & Layout
Sheila Walsh

Circulation
(clockwise)
Brenda Twersky

Circulation
(counterclockwise)
James Mahoney

EDITORIAL BOARD

Allergy and Immunology
Max Samter, M.D.
Grant Hospital of Chicago
Chicago, IL

Anthropology
Prof. Jonathan Marks
Yale University

Astronomy
Jay M. Pasachoff
Hopkins Observatory
Williams College

Robert Kirshner
Harvard University

David Slavsky
Loyola University of Chicago

Biochemistry
Dr. Max E. Rafelson, Jr.
Presbyterian–St. Luke's Hospital
Chicago, IL

Biomaterials
Alan S. Litsky
Ohio State University

Biophysics
Leonard X. Finegold
Drexel University

Bureaucratic Affairs
Miriam Bloom, Ph.D.
U.S. Consumer Safety Commission
Jackson, MS

Chemistry
Dudley Herschbach°
Harvard University

Richard E. Jensen, Ph.D.
Forensic Associates, Inc.
Minneapolis, MN

Benjamin J. Luberoff, Ph.D.
CHEMTECH
Summit, NJ

Linus Pauling°
The Linus Pauling Institute
Palo Alto, CA

Computer Science
Prof. Dennis J. Frailey
Texas Instruments
Plano, TX

Computer Security
Dr. Harold J. Highland
Ed.-in-Chief, Comp. & Sec.
Elmont, NY

Computer Software
Heidi Roizen
T/Maker Company
Mountain View, CA

Dentistry
Walter J. Kent, D.D.S.
Wyckoff, NJ

Joseph J. Marbach, D.D.S.
Dir., Facial Pain Clinic
New York, NY

Developmental Psychology
Neil V. Salkind
University of Kansas

Ecology and Evolution
Prof. Leigh Van Valen
University of Chicago

Economics
Ernst W. Stromsdorfer
Washington State Univ.

Prof. Hein Schreuder
Univ. of Limburg
Maasricht
The Netherlands

Engineering
Ross F. Firestone
IIT Research Inst.
Chicago, IL

Forestry
Prof. R.W. Stark
University of Idaho

Geography
G.H. Dury, D.Sc.
Risby, Bury St. Edmunds
Suffolk, England

Geology
J. Russell Boulding
Bloomington, IN

John C. Holden
Omak, WA

John F. Splettstoesser
Polar Consulting, Inc.
Rockland, ME

History of Science and Medicine
T. Healey, M.D., Ch.B.
Bamsley, Yorkshire, England

Informational Science
Dr. Norman D. Stevens
Univ. of Connecticut

Intelligence
Marilyn vos Savant
New York, NY

Law
Ronald A. May
Wright, Lindsey & Jennings
Little Rock, AR

Management
Dr. Waino W. Suojanen
Georgia State Univ.

Marine Geology
Dr. Constance Sancetta
Lamont-Doherty Geo. Observatory
Columbia University

Materials Science
Robert M. Rose, Sc.D., P.Eng.
MIT

Mathematics
Dr. Lee Segel
Weizmann Inst.
Rehovot, Israel

Medicine and Law
Michael Applebaum, M.D., J.D.
Medical & Legal Services
Chicago, IL

Methodology
Dr. Stanley Rudin
Lima, OH

Rodney L. Levine, M.D.
NIH
Bethesda, MD

Microbiology
Prof. G. Roland Vela
North Texas State Univ.

Molecular Biology
Sir John Kendrew°
Cambridge, England

Neuroengineering
Jerome Y. Lettvin
Rutgers University

Nursing
Elizabeth W. Riggs
DeKalb County Board
of Health
Decatur, GA

Opthalmology
Pinar Aydin, M.D.
Hacettepe University
Ankara, Turkey

Orthopedic Surgery
Glenn R. Johnson, M.D.
Bemidji, MN

Pediatrics
Ronald B. Mack, M.D.
Bowman Gray School of Medicine

Robert E. Merrill, M.D.
Salado, TX

Pharmacology
G. Kimmell Stanton
Dir., Pharmacy Services
Norman Regional Hospital
Norman, OK

Physics
Sheldon Lee Glashow°
Harvard University

Leon Lederman°
University of Chicago

Dr. H.J. Lipkin
Weizmann Inst.
Rehovot, Israel

Thomas G. Kyle, Ph.D.
Los Alamos National Lab.
Los Alamos, NM

Mel Schwartz°
Brookhaven National Lab.
Upton, N.Y.

Political Science
Dr. Edward Mickolus
Vinyard Software
Falls Church, VA

Prof. Richard G. Neimi°°°
Univ. of Rochester

Psychiatry and Neurology
Robert S. Hoffman, M.D.
Peninsula Neurol. Assoc.
Daly City, CA

Psychology
Louis Lippman
Western Washington Univ.

Pulmonary Medicine
Dr. Traian Mihaescu
Clinic of Pulmonary Disease
Iasi, Romania

Radiology
David N. Rabin, M.D.
Highland Park Hospital
Highland Park, IL

Science and Public Policy
Daniel S. Greenberg
Washington, DC

Stochastic Processes
Dr. George August
Silver Spring, MD

Surgery
Douglas Lindsey, M.D.
Univ. of Arizona Med. Ctr.

Urology
Mark Sullivan, M.D.
Mission Viejo, CA

Veterinary Medicine
J.F. Smithcors
Santa Barbara, CA

°Nobel Laureate
°°Convicted Felon
°°°Misspelled

CONTENTS

CONTENTS

INTRODUCTION AND APOLOGY

Marc Abrahams
Editor, The Journal of Irreproducible Results

"Science in general can be considered a technique with which fallible men try to outwit their own human propensities to fear the truth, to avoid it, to distort it."

—Psychologist A.H. Maslow

Scientists are curious creatures.[1] They have an incurable itch to understand EVERYTHING. They look at things, and poke at things, and then they write reports about what they have found. Their colleagues read these reports, and attempt to duplicate what the original scientists did and saw.

Newly discovered phenomena can be hard to pin down. Some are very rare. Some are difficult to see. Some turn out to be misleading chance occurrences. Some turn out to be just wishful thinking or self-delusion on the part of the scientist. If other people cannot reliably reproduce what the original researcher found, having looked for it in exactly the same way, the phenomenon is said to be *irreproducible*.

Many new discoveries at first seem to be irreproducible, and are for a time the subject of scorn and the butt of jokes. Only through the dogged, determined efforts of curious individuals do new ideas come to be examined, tested, and sometimes accepted. The fact that something seems irreproducible doesn't necessarily mean it is illusionary. It just means that we shouldn't accept the phenomenon as real without doing further investigation.

Most science journals[2] are hesitant to print reports about irreproducible phenomena. The *Journal of Irreproducible Results* prints nothing but. This book, and the earlier collection entitled *The Best of the Journal of Irreproducible Results*, gathers some of the most provocative, and most and least important, research performed during the last 37 years.

The reports included here come from many countries. The researchers come from a variety of backgrounds. They range from a Nobel Laureate (Dudley Herschbach, author of "Sweet Seventeen") to an internationally acclaimed jazz harpist (Deborah Henson-Conant, curator of the Museum of Burnt Food).

The topics include nearly everything under the sun, and much that goes beyond that complex body and/or inside it (see, for example, "A Briefer History of the Universe"). There is much here to stimulate the mind, the libido, and the digestion. Burnt toast (A Decade of Burnt Food), erotic art ("The Mappeltree Science Art Controversy"), the television show *I Love Lucy* ("Titular Dominance in *I Love Lucy*"), coffee-brewing rituals ("Nobel Thoughts: Eric Chivian"), zippers ("More About Zippers"), Abraham Lincoln's facial hair ("Abraham Lincoln's Mustache"), muscular frogs ("The Effects of Anabolic Steroids on Northern Grass Frogs"), literary fashion ("In Memorium: Raul de Womynn, The Father of Desensification"), "Flashbulb Photography at Niagara Falls," "Infectious Diseases in Bricks," and more. There is also, of course, a profusion of research about that most exotic and provocative of recent scientific ideas: cold fusion. An entire section is devoted to cold fusion, as are several items in the various "Scientific Gossip" columns.

From its beginning, *JIR* has also conducted guerilla warfare against unnecessary mindnumbing jargon, petty academic politics, and self-important bureaucrats. Our weapons of choice have always been subtlety and blunt instruments.

The title of this book, *Sex As a Heap of Malfunctioning Rubble*,[3] is loosely taken from the "Elegant Results" column that appears on page 121. The column, a regular feature of *JIR*, is, we believe, unique in the annals of science. Alice Shirell Kaswell collects important scientific discoveries that appear in overlooked research journals such as *Cosmopolitan, Vogue, Ladies' Home Journal, Golf, GQ, and Modern Bride*. Science is a vast and wide-ranging pursuit.

"Elegant Results" extends a traditional service for which *JIR* is famous. The "JIR Recommends" column alerts our readers to groundbreaking research papers that have been published in other science and medical journals. A few examples: "Elucidation of chemical compounds responsible for foot malodour;" "Effects of drinking hot water, cold water, and chicken soup on nasal mucus velocity and nasal airflow resistance;" and "Termination of intractable hiccups with digital rectal massage." Classic works, all of them, and available in any good science or medical library.

Each year *JIR* is involved in mounting the Ig Nobel Prize Ceremony, in which genuine Nobel Laureates present Ig Nobel Prizes to individuals who have notably irreproducible achievements in science and other areas of human endeavor. The first prizes were awarded privately in 1968. Reports on the first two public ceremonies begin on page 168. The ceremony is always held at MIT on the first Thursday in October. You are cordially invited to attend.

Your Role

We hope that this book will inspire you to conduct your own research on a topic that obsesses you. Many of the research reports in this book were submitted by brave investigators not unlike yourself. The "Information for Contributors" (see page 181) section can guide you in your efforts.

Apology

During the past few years, many people have written to *JIR* complaining about typographic errors. They used to look forward to counting the typos in each new issue, but now the number has dropped from a dependable 40 to 50 per issue to five or even fewer. The change is due to an oversight on our part. We hereby apologize for it.

Thanks

I want to give special thanks to several people who have played important and colorful roles in *JIR*'s recent history: Alex Kohn; George Scherr; Jim Krosschell; Sheila Walsh; Stephen Drew; Alice Shirell Kaswell; our intrepid associate editors, Michèle Meagher, Mark Dionne, Bertha Vanatian, and Stanley Eigen, mathematical raconteur of raconteurs; Norman Stephens of the Molesworth Institute; Kathy Thurston, Warren Seamans and the amazing staff at the MIT Museum; the astounding *JIR* editorial board members who devote so much time, energy and wit; Eric Chivian, Jerry Friedman, Sheldon Glashow, Dudley Herschbach, Henry Kendall, Mel Schwartz, the five fearless and mischievous Nobellians who made the First Annual Ig Nobel Prize Ceremony such a success, along with their collaborator Karen Hopkin, the Angry Woman in the Red Dress; Gina Maniscalco; Sid Abrahams; Roland Sharrillo; Lois Malone; Suzanne Marshall; Dr. Robert Richard Smith; Sheldon Cohen of Out of Town News; Michael Harris of Daybreak Distributors; Mary Chung; Bruce Gellerman; Dan Quayle; John Chou; Astrid Hiemer; Mark Fitzgerald of ISI; Clockwise Twersky and Counterclockwise Mahoney. My warmest thanks also to my editor Suzanne Rafer, and her associate on this project, Jim Gorman. Thanks, too, to designer Lori S. Malkin. And a special and continuing thank you, thank you, thank you, thank you, thank you, thank you, thank you, thank you, thank you, thank you, thank you, thank you, thank you, thank you to Martin Gardner.

---

1. See all definitions of the word *curious*.

2. There are now approximately 10,000 science journals being published throughout the world. Every scientific specialty also produces a host of smaller bulletins and newsletters. All told, there are somewhere between 80,000 and 100,000 scientific publications being produced on a regular basis. A few, such as **Nature** and **Science** are known to all scientists and to much of the general public. Others, such as the **Journal of Gastroenterology Motility Studies**, are not so well known, and are seen only by members of very specialized communities. How many research reports appear in these journals? The Institute for Scientific Information estimates that the leading 10,000 science journals produce *approximately 1.8 million articles each year.* Add in the number of articles appearing in the other publications, and the number is somewhat more than staggering.

3. The title of this book, *Sex As a Heap of Malfunctioning Rubble,* is also part of an experiment being conducted by the publisher. The book is also being published under another title: *A Comprehensive Compendium and Detailed Descriptions of Experiments and Other Research Activities That Yield Results of an Apparently Irreproducible Character.* The experiment is this: Does the presence of the word "sex" in a book's title cause people to pick up the book? We eagerly await the results of any experiment you yourself might conduct on this matter.

THE BEGINNINGS OF JIR

Alexander Kohn, Ph.D.
Founder and Editor Emeritus

The *Journal of Irreproducible Results (JIR)* was born in 1956. In 1951, as a fresh-baked Ph.D., I joined a small group of Israeli scientists to start a new research institute, the Israel Institute for Biological Research in Ness Ziona. The story of *JIR* begins on a bus trip to the Dead Sea, organized by the University for its teachers in 1956. During the long ride, a discussion started up as to why so much glassware had been mysteriously disappearing in the chemical and microbiological laboratories. Jokes proliferated and it occurred to me that the problem of inactivation of glassware in these laboratories could well be described in scientific terms using practical examples. At the end of the trip I composed a story entitled "The Kinetics of Inactivation of Glassware." I made a number of copies and sent them to the trip participants; as a reference, I called the article Vol. 2, No. 1 of *JIR*. The reactions to this new journal were quite encouraging and there was a persistent demand to continue with such a publication.

I learned that Professor Harry J. Lipkin, a physicist at the Weizmann Institute in Israel, had been slyly writing some irreproducible humorous articles for his own and his friends' enjoyment. I invited him to join me as co-editor of *JIR*, and our collaboration lasted for about a decade. We used a stencil to make a master copy and printed as many copies as there were interested readers. We charged $1.00 per year for two issues. When the number of subscribers had rocketed to 800, our effort to be the editors as well as the publishers of the *Journal* became impossible to sustain. (After all, we had our "serious" work as scientists.)

The publication of the *Journal* was taken over in 1963 by Dr. George H. Scherr (Chicago Heights, Illinois). For a number of years we used to send him a master copy of the journal which he then printed and distributed. Later, the selection of articles was made upon the recommendation of a rising number of associate editors (their number reached about 40 in 1990). Scherr sold *JIR* to Blackwell Scientific Publishers in

Cambridge, Massachusetts in 1990, and it is now edited by Marc Abrahams.

To return to our history: Issue No. 2 of the *Journal* was devoted to the problem of zippers. Lipkin wrote a dissertation, entitled "Theoretical Zipperdynamics" where he discussed the Zipperbewegung, the semi-infinite and the finite Zipper, relying heavily on the Schroedzipper equation. He hoped that the exploitation of zipperic energy would produce some useless research. As I shall describe later, it indeed did. I myself contributed an article concerning the applications and complications of zippery mechanisms in biology and medicine. I based some of my explanations on the (recent at that time) theory of Watson and Crick about the zipperlike replication of DNA. I suggested some practical applications for zippers in medicine, such as their use by surgeons who had forgotten some surgical instruments in the internal cavities of patients they had operated on. Amazingly, this "science fiction" prediction became a reality after some 20 years. I also suggested development of an electronic, remote control device for opening and closing a zipper installed on the mouth of

Alexander Kohn, Ph.D.

a too-talkative spouse. Unfortunately, this has not yet materialized.

A very unexpected development relating to Lipkin's article came about 20 years after its publication. A scientist from Tuebingen, Germany phoned me asking for an exact reference to Zipperdynamics. He was especially interested in the Schroedzipper equation. He explained that he needed this information for his Ph.D. thesis which he had to submit the next day. Some weeks later I received a copy of his thesis with the impossible title, containing a noun of 50 letters: "Aufbau und Anwendung eines Fouriertransformkernquadrupolresonanz-spectrometers." In it he indeed quoted in all seriousness Lipkin's paper in *JIR* and even attached a photocopy of that article to the thesis.

Volume 3 of *JIR* started with a paper by a professor of the Weizmann Institute writing under the pseudonym of S.U. Perkinsey: "The Chemistry of Copulation, VI: The Interpretation in Terms of the Activated State Theory of Chemical Kinetics." The paper was largely misunderstood and we received many protests against it. In the experimental part, it described measurements and calculations of bond length, the excitation and seduction potentials, resistance and energy of activation, etc.

With the passage of time, the number of issues per year steadily increased to the present six (and so did the price!). The journal became the official organ of the Society for Basic Irreproducible Research (with its infamous secretary X. Perry Mental). In due course it dealt with researchmanship (i.e., the art of conducting and publishing research without actually doing it); obscurantism (i.e., the use of ununderstandable jargon in scientific publications); development of new scientific laws (e.g., Gordon's law: If an experiment is not worth doing, it is not worth doing well); and genuine but funny quotations from the scientific literature. (Some of them appeared later in a booklet called *Don't Quote Me.*)

In its heyday, the journal reached some 40,000 subscribers in 52 countries. The number of contributions to *JIR* exceeded its capacity (32 pages per issue) and it became the journal in the scientific community with the highest rejection rate (about 70%). For example, we rejected a paper by Nicolau and Perrin, entitled "The Reconstitution of Spoiled Béarnaise Sauce," which was later published in **Nature**.

JIR documented the invention of some new chemicals, devices, and methods. One of them was to supply semi-arid countries in the Middle East or South America with water melted from towed-in icebergs. Icebergs from Antarctica were to be towed to these countries and deposited in artificial lakes. Some 15 years after our suggestion, the matter was seriously taken up and supply of water to Saudi Arabia by this method was earnestly considered.

One of *JIR*'s outstanding contributions to science was the invention of the contraceptive, No-Acetol, a six-membered ring of nitrogen and oxygen, constructed so that it had a NO in every position.

To end the story of the first years of *JIR,* here is an amusing (absolutely true) story, told to *JIR* by Dr. Ralph A. Lewin of the Oceanographic Institute in La Jolla, California:

When we first moved to La Jolla and began to settle down, we needed curtains among other things. We found a suitable set of draperies at JC Penney's, a large department store in San Diego. The saleswoman was reluctant to accept a personal check in payment unless we could establish our *bona fides.* I offered her my driver's license—sorry, not acceptable; my United States Immigration Card I.151—not acceptable; my Social Security card—not acceptable. In desperation I handed her my wallet with little else but cards, and asked her to see if she could find one which would assuage her doubts as to our responsibility. She thumbed past memberships in the American Association of University Professors, The International Esperanto Society, the San Diego Zoological Society, and finally stopped at the Identification Card No. 576 of the Society for Basic Irreproducible Research, signed on February 30 1960 by no less a personage than X. Perry Mental himself. "This will do," said the discriminating lady with a satisfied smile. She accepted my check, copied the details of my *SBIR* membership on its back and slipped it under the cash register. She packaged the curtains and bade us a sunny "good afternoon."

Well, there is absolutely nothing like science.

CHAPTER 1

THE SECRETS OF THE UNIVERSE

A BRIEFER HISTORY OF TIME

Michèle M. Meagher
Wisdom Simulators, Inc., Cambridge, Massachusetts

BANG!

HEAVIEST ELEMENT DISCOVERED

Thomas G. Kyle
Los Alamos, New Mexico

The heaviest element known to science was recently discovered at a prominent national laboratory. The element, tentatively named administratium (Ad), has no electrons or protons, thus having atomic number zero. It does, however, have one neutron, 75 associate neutrons, 125 deputy associate neutrons, and 11 assistant deputy associate neutrons. This gives it an atomic mass of 312. The 312 particles are held together in the nucleus by a force that involves the continuous exchange of mesonlike particles called memoöns.

Since it has no electrons, administratium is inert. Nevertheless, it can be detected chemically because it seems to impede every reaction in which it takes part. According to Dr. M. Languor, one of the discoverers of the element, a very small amount of administratium caused one reaction that normally occurs in less than a second to require over 4 days to go to completion.

Administratium has a half-life of approximately 3 years, at which time it does not actually decay. Instead, it undergoes an internal reorganization in which associates to the neutron, deputy associates to the neutron, and assistant deputy associates to the neutron all exchange places. A tendency has been observed for the atomic mass to actually increase during each reorganization.

MAGNETITE SPHERULE GOLF BALL

The "golf ball" is actually a magnetite spherule. The "tee" is a crystal of $CaSO_4$. Both occur naturally in bentonite clay. This scanning electron micrograph was taken by Fraser King of Whiteshell Laboratories, Pinawa, Manitoba, Canada.

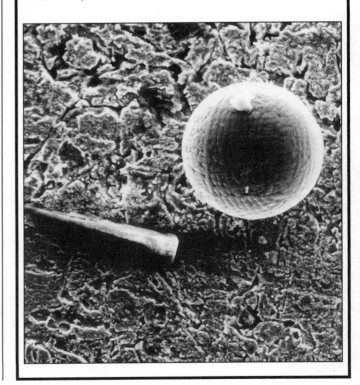

THE BLACK HOLE OF CALCUTTA

Jonathan Small, Mahomet Singh, Abdullah Khan, and Dost Akbar

University of Caerphilly, Caerphilly, Wales

Abstract

An unusual box was discovered in Calcutta, India. Evidence is presented that it contains a black hole. The physics that underlie this phenomenon are currently not well understood.

Discovery of the Box

In 1990, while engaged in an archeological dig near the city of Calcutta, the authors uncovered a box of unusual design. The box was found to exhibit peculiar properties. It was removed to the physics laboratory at the University of Caerphilly, where one of the authors (Small) submitted it to extensive analysis.

Historical Background

While vague references to mysterious objects exist in much of India's copious historical literature, the authors have not been able to identify any specific references to this box or to boxes of this general nature.

Left to right: *Dr. Khan, Dr. Small, and Dr. Singh hold the box containing the black hole of Calcutta moments after its discovery.(Not pictured: Dost Akbar.) Photo: Dr. Robert Richard Smith/Minimum Wage Art.*

Properties of a Black Hole

A black hole is an object whose mass is extremely compressed. Its gravitational force is so strong that nothing, not even light, can escape it. However, in the mid-1970s, Stephen Hawking of Oxford University showed that, under the laws of quantum mechanics, black holes must emit some radiation. A black hole exhibits certain specific characteristics. These characteristics are described in detail in Hawking (1975) and Penrose (1979).

Properties of the Box

The box is made of wood. The dimensions are approximately 10 cm x 10 cm x 15 cm. The box emits steady, extremely low-level radiation in all measurable parts of the spectrum. This radiation presents all the characteristics of the Hawking-Penrose theoretic model of a black hole.

Discussion

Because a black hole sucks in all planets, stars, and light that come near it, it was believed that black holes exist only in distant reaches of space far from the earth. This black hole, however, was found in a wooden box in India.

This fact raises two immediate questions:

1. Why hasn't this black hole sucked the earth, the sun, and the entire solar system into itself?
2. This box containing this black hole must be unusually strong. Was it constructed using standard engineering techniques?

The authors have no satisfactory answer for either of these questions. We invite suggestions and comment.

Addendum

Despite their best efforts, the authors have thus far been unable to open the box.

REFERENCES

Beckenstein J. Black holes and entropy. *Phys Rev* 1972; D7:2333–2346.

Hawking SW. Particle creation by black holes. *Commun Math Phys* 1975; 43:199–220.

Penrose R. Singularities and time-asymmetry. In: Hawking SW, Israel W (eds): *General Relativity: An Einstein Centenary.* Cambridge University Press, 1979.

Shackleton B, Schayek, LL. 10 Fascinating sealed boxes. In: Wallechnisky D, Wallace I (eds): *The People's Almanac #2.* Bantam: New York; 1978, 1108–1110.

APOLOGY

Because of an oversupply of red tape, the Cosmic™ all-natural superstrings advertised in *JIR* 35:7 are not available.

The U.S. Postal Service (USPS) has ruled that superstrings violate its size requirements. According to the USPS, superstrings are both too small to be insurable against loss and too massive to be carried in standard packaging. Moreover, the postal carriers union forbids its members from carrying 11-dimensional objects.

Full refunds will be sent to everyone who placed an order. We apologize for any inconvenience or disappointment this may cause our readers.

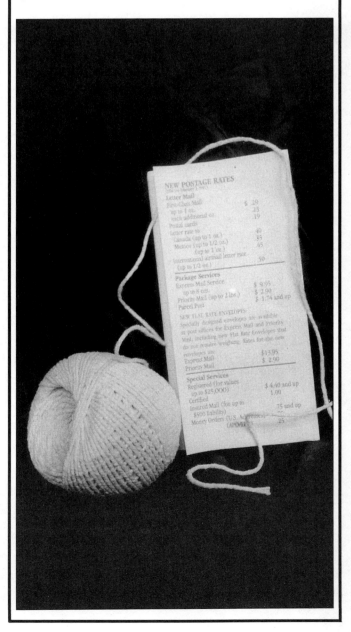

A CALL FOR MORE SCIENTIFIC TRUTH IN PRODUCT WARNING LABELS

Susan Hewitt and Edward Subitzky
New York, New York

As scientists and concerned citizens, we applaud the recent trend towards legislation that requires the prominent placement of warnings on products that present hazards to the general public. Yet we must also offer the cautionary thought that such warnings, however well-intentioned, merely scratch the surface of what is really necessary in this important area. This is especially true in light of the findings of 20th century physics.

We are therefore proposing that, as responsible scientists, we push for new laws that will mandate the conspicuous placement of suitably informative warnings on the packaging of every product in every category offered for sale in the United States of America. Our suggested list of required warnings appears below.

WARNING: This Product Warps Space and Time in Its Vicinity.

WARNING: This Product Attracts Every Other Piece of Matter in the Universe, Including the Products of Other Manufacturers, with a Force Proportional to the Product of the Masses and Inversely Proportional to the Distance Between Them.

CAUTION: The Mass of This Product Contains the Energy Equivalent of 85 Million Tons of TNT per Net Ounce of Weight.

HANDLE WITH EXTREME CARE: This Product Contains Minute Electrically Charged Particles Moving at Velocities in Excess of 500,000,000 Miles per Hour.

CONSUMER NOTICE: Because of the "Uncertainty Principle," It Is Impossible for the Consumer to Find Out at the Same Time Both Precisely Where This Product Is and How Fast It Is Moving.

ADVISORY: There is an Extremely Small but Non-zero Chance That, Through a Process Known as "Tunneling," This Product May Spontaneously Disappear from Its Present Location and Reappear at Any Random Place in the Universe, Including Your Neighbor's Domicile. The Manufacturer Will Not Be Responsible for Any Damages or Inconvenience That May Result.

READ THIS BEFORE OPENING PACKAGE: According to Certain Suggested Versions of a Grand Unified Theory, the Primary Particles Constituting This Product May Decay to Nothingness Within the Next 400,000,000 Years.

THIS IS A 100% MATTER PRODUCT: In the Unlikely Event That This Merchandise Should Contact Antimatter in Any Form, a Catastrophic Explosion Will Result.

PUBLIC NOTICE AS REQUIRED BY LAW: Any Use of This Product, in Any Manner Whatsoever, Will Increase the Amount of Disorder in the Universe. Although No Liability Is Implied Herein, the Consumer Is Warned That This Process Will Ultimately Lead to the Heat Death of the Universe.

NOTE: The Most Fundamental Particles in This Product Are Held Together by a "Gluing" Force About Which Little Is Currently Known and Whose Adhesive Power Can Therefore Not Be Permanently Guaranteed.

ATTENTION: Despite Any Other Listing of Product Contents Found Hereon, the Consumer Is Advised That, in Actuality, This Product Consists Of 99.9999999999% Empty Space.

NEW GRAND UNIFIED THEORY DISCLAIMER: The Manufacturer May Technically Be Entitled to Claim That This Product Is Ten-Dimensional. However, the Consumer Is Reminded That This Confers No Legal Rights Above and Beyond Those Applicable to Three-Dimensional Objects, Since the Seven New Dimensions Are "Rolled Up" into Such a Small "Area" That They Cannot Be Detected.

PLEASE NOTE: Some Quantum Physics Theories Suggest That When the Consumer Is Not Directly Observing This Product, It May Cease to Exist or Will Exist Only in a Vague and Undetermined State.

COMPONENT EQUIVALENCY NOTICE: The Subatomic Particles (Electrons, Protons, etc.) Comprising This Product Are Exactly the Same in Every Measurable Respect as Those Used in the Products of Other Manufacturers, and No Claim to the Contrary May Legitimately Be Expressed or Implied.

IMPORTANT NOTICE TO PURCHASERS: The Entire Physical Universe, Including This Product, May One Day Collapse Back into an Infinitesimally Small Space. Should Another Universe Subsequently Reemerge, the Existence of This Product in That Universe Cannot Be Guaranteed.

NOBEL THOUGHTS

Profound insights of the laureates

Marc Abrahams

Sheldon Lee Glashow is Higgins Professor of Physics and Mellon Professor of the Sciences at Harvard University. He was awarded the Nobel Prize in Physics in 1979 for his work leading to development of the electroweak theory, uniting electromagnetism and the weak nuclear force, two of the fundamental forces of nature, under a single gauge theory. We spoke in his office in Cambridge, Massachusetts.

What time to you get up in the morning?
About 8 A.M.

What time do you arrive at the lab?
About 9:00. Sometimes I get there as early as 8:30, sometimes as late as 10:00.

How do you get to work?
I take the Subaru, unless it doesn't start. In that case, I stay home.

Do you have any advice for young people who are entering the field?
I don't remember meeting any.

Contains 100% gossip from concentrate

Compiled by Stephen Drew

Long-Term Cosmetics Look Good in Monkeys

Research with biocompatible polymers and collagen adhesives hints at a new generation of super-extended-wear mascaras, lipsticks, and eye shadow. Researchers are already fitting monkeys with these "almost forever" cosmetics, which humans may someday wear day and night without the risks of skin damage, infection, or "running" that are inherent in today's cosmetic products.

Parnell Hincher of the Helmsley Beauty Research Institute in Binghamton, New York, is experimenting with a technique called microsphere unguent surgeoplasty. It involves scraping away the thin outer layer of epithelial cells covering the cheeks, lips, and nose, then attaching a refractive, biocompatible covering, or lenticle. The lenticle is made of chemically altered collagen (a naturally occurring protein), dextrose, sucrose, polystyrene, and a quasicrystal form of a substance known as Stingo-12, which is derived from the pheromones produced by several species of Ecuadoran tree ants. Hincher says that the colors can be repeatedly altered by modifying the monkeys' diet. He predicts that human trials will begin soon.

Multi-Religion Groups Seen Emerging

Dramatic changes may be coming to the religious and political landscape of Western countries. The recent emergence of "multi-religions" such as Jews for Jesus, Methodists for Mohammed, and the Southern-based Baptists for Buddha may herald a period of drastic societal realignment. The role of smaller, less influential groups (Episcopalians for Elvis, Atheists for Adlai, Mormons for Marx, Krishnas for Khomeini, etc.), which have fewer than 10,000 members, is seen as harder to predict.

SLEEP RESEARCH UPDATE

- SD has stopped sleeping with NF and is now sleeping with TL.
- NF now sleeps with FP and LB.
- TdeC is now sleeping with RR.
- Dr. NM is sleeping with PL, BR, NS, EK, and Dr. CL.
- PN is sleeping alone.
- GY now washes her feet regularly and is sleeping with FP.

A Driller's Nightmare May Be a Dream Food Source

Drillers for oil and natural gas have long feared a snowball-like mass of water and hydrocarbons called clathrate hydrate. This "ice," which forms at temperatures above the freezing point of water, can jam valves that stop the upward flow of gas, allowing the gas to rush up a bore hole and explode.

But this slushy villain is also an abundant food source. The biomass that is preserved in hydrate ice deposits on the sea floor and in the Arctic could hold considerable quantities of commercially recoverable foodstuffs. Its

temperatures and high pressures. Kates and other researchers are attempting to remove the pungent odor that is imparted by the methane.

Healthier Cigarettes

Several major tobacco companies are racing to bring to market cigarettes that they claim will help prevent heart attacks. The cigarettes, which will be laced with aspirin, will be test marketed to several groups. According to industry sources, the first such group will be pregnant women. "It's a woman's right to protect herself and her unborn child against heart disease," reads the headline in a full-color advertisement scheduled to make its first appearance in the upcoming issues of several women's magazines.

Puzzling Venus Features Explained

The curious surface patterns visible in recent photographs of Venus have been explained.

Historic Evidence May Alter Calendar

A team of Conway University anthropologists, linguists, and archeologists has found evidence that the day we now refer to as "Thursday" is actually Wednesday, and that the day known to us as "Wednesday" is really Thursday. The research team, led by Margaretha Gimptle-Ross, Bizi Yu, and Hans Salzman, bases its contention on records they recovered from a dig site in the ancient city of Ur.

nutritional caloric content could surpass that of the world's total stocks of industrial-grade fish and kelp, according to an analysis performed by Jesse Kates of the Eureka College Agronomy Department. The hydrates apparently form when organic matter decays, releasing methane gas. The methane recombines with the partially decayed foodlike particles, which are then trapped in "cages" of water molecules created under low

Fred Winkowski

Discounts Stir Controversy

A science bookstore's new discount policy is causing passions to run high among scientists, politicians, and religious authorities in Oregon. New Age Alchemy Used Books has placed a sign in its window offering "Discounts For The Unborn."

U.S. Election Proposal Is Seen Gaining Support

There appears to be growing support in the United States, especially among retired politicians, for a proposed constitutional election reform. Under the proposal, anyone who wishes to run for elected office would be prohibited from doing so. The proposed constitutional amendment cites a need to "have a higher caliber of persons enter political life." It reportedly has backing from all the living former U.S. Presidents except Ronald Reagan.

Progress on Coal Fusion

Researchers at universities in Japan, Switzerland, and the United States are all claiming credit for a technique that could be used to fuse chunks of low-grade coal into an inexpensive, highly compact, yet extremely potent, new form of cattle feed. The new feed is called "diamondane" because of its misleading physical resemblance to the precious mineral.

LETTERS TO THE EDITOR

The Ongoing Search for Truth and Knowledges

*Editor's Note: As **JIR** goes to press, we have not yet received a reply to this inquiry:*

October 24, 1990

Editor
National Enquirer
600 South East Coast Avenue
Lantana, Florida 22464

Dear Editor:
Do you have any articles which you cannot use because they are true? If so, may we print them?

I look forward to hearing from you.

Sincerely,

Stephen Drew

Stephen Drew
Assistant Editor
Journal of Irreproducible Results

CHAPTER 2

SCIENCE ART CONTROVERSIES

FLASHBULB PHOTOGRAPHY AT NIAGARA FALLS

Robert L. and Susan E. Feldman
Ithaca, New York

Introduction

On the first night of the authors' honeymoon at Niagara Falls, we strolled over to Victoria Park on the Canadian side to romance by the Falls. Because of a strike by the park workers, the floodlights that normally would illuminate the Falls were off. Since there was a new moon, the Falls were almost completely dark. As we stood there, we were intrigued by the many visitors who were taking photographs of the Falls using flashbulbs. Every few seconds, a flashbulb went off somewhere. The authors' research (Feldman and Goodman, 1988) indicates that a normal flashbulb has no effect on an object more than 30 feet from the camera, so we were puzzled by this phenomenon. Scientific curiosity got the better of us, and we decided to devote our honeymoon to an investigation into the success rate of these photographers. (Although it did not occur to us at the time, this also made our honeymoon tax-deductible.)

Methodology

We selected the first 100 tourists whom we observed taking flash photographs of the Falls and asked for each subject's name, mailing address, and phone number. We repeated this process at the same time on each of the next four nights, for a total sample size of 500.

As soon as we returned home, we designed a questionnaire to send to our subjects. We mailed it 1 month later, with two follow-up waves to nonrespondents at 3-week intervals. Finally, we made threatening phone calls at odd hours to the few remaining nonrespondents. We were thus able to achieve a 100% return rate.

The questionnaire and results are shown in Table 1.

Interpretation

The high percentage of "not visible at all" photos (95%) supports the authors' hypothesis that a flashbulb has too short a range to light Niagara Falls. What is surprising is that some of the photographs (albeit a small percentage) were rated as "easily visible" or "visible with difficulty." However, in a sample size of 500 it is quite possible that two tourists pressed the shutter release at approximately the same moment so that the light from both flashbulbs added and was enough to dimly light at least part of the Falls. This is the so-called "multiple simultaneous flash phenomenon," or MSFP (Hypo, 1977).

In the one case rated "easily visible" we hypothesize that three or more flashbulbs went off simultaneously. However, since this respondent also rated himself as an "incompetent" photographer, we view his opinion as suspect. The fact that only the Horseshoe Falls showed the MSFP is also logical, since they are closer to the Canadian side than the American Falls.

There was a distinctly higher incidence of successful photographs on the fourth night of our study. On this night it was raining. At first blush this finding seemed counterintuitive to us, but then we recalled that there was also quite frequent thunder and lightning. (In fact, we had to exclude one subject who was struck by lightning and substitute another.) At any rate, the successful photos were probably due not to the flashbulb but to illumination from the lightning.

Perhaps the most surprising result is the high percentage (78%) of respondents who indicated that they would take flash photographs at night again at Niagara Falls. One possible explanation was suggested by Bluedot (1982). He surveyed tourists taking flash pictures of a lunar eclipse in Antarctica and found that 82% of those surveyed *expected* their nighttime photographs to be dark, since, after all, that's what night is all about. So rather than being upset by rolls of completely black pictures, tourists may be pleased by this capture of expected reality.

Table 1

1. How many nighttime flash photographs did you take at Niagara Falls? *Total for all respondents:* 76,970 photographs.

In how many of these were Niagara Falls:

Easily visible?	1 photograph
How many were of the American Falls?	0
How many were of the Horseshoe Falls?	1
Visible with difficulty?	9 photographs
How many were of the American Falls?	0
How many were of the Horseshoe Falls?	9
Not visible at all?	72,880 photographs (see Figure 1)
Other—please explain:	4,080 (equals 180 rolls)
Not developed yet:	148 rolls
Forgot the film was still in the camera:	22 rolls
Forgot to put film in the camera:	7 rolls
Used flash unit without camera:	1 roll
Camera fell over the Falls:	1 roll
Film confiscated by border guards:	1 roll

2. How would you rank yourself as a photographer?

10	Professional
128	Seasoned amateur
348	Amateur
2	Incompetent
4	Not sure
8	No reply

3. If you were to visit Niagara Falls again, would you plan to take flash photographs again?

390	Yes
76	No
26	Not sure
8	No reply

Another possible explanation of satisfaction with a 95% failure rate with nighttime flash pictures is found in survey results by Triepodd (1986). She obtained the names and addresses of a sample of visitors taking daytime pictures at Disneyland and interviewed them by phone 3 weeks later. She found that 95% of their *daytime* photographs did not come out. So perhaps this failure rate is considered normal by tourists, whether it is day or night.

Future Research

An exciting experiment is being carried out by the authors' equipment that was recently launched along with the Hubble space telescope. We have installed a camera with flash unit, which will attempt to take flash pictures of black holes through the telescope.

Over the next year we intend to apply our methodology to flash photographs taken (a) from the Empire State Building, (b) in New York's Hayden Planetarium, and (c) at the Grand Canyon (our second honeymoon, also tax-deductible). We also plan on returning to Niagara Falls to study tourists taking flashbulb photographs during the daytime.

Figure 1. Typical flash photos of Niagra Falls.

REFERENCES

Bluedot F. Flash Photography During a Lunar Eclipse. *Jrnl Low Temp Photog* February 1982.

Feldman R, Goodman S. Tripping the Light Fantastic. *Read Dig* August 1988.

Hypo G. A Random Walk with Flashbulbs. *Jrnl of Irrepressible Results,* December 1977.

Triepodd S. The Mickey Syndrome. *Disney Rev* October 1986.

THE MAPPELTREE SCIENCE ART CONTROVERSY

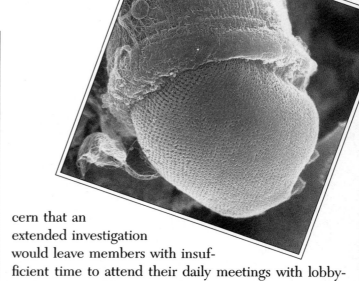

Bruce Gellerman

WBUR, National Public Radio, Boston, Massachusetts

U.S. Senator Johnny Holmes's move to bar Federal funding of "obscene and indecent science art" is sending shock waves through both the scientific and the art communities.

Holmes Is Where the Art Is

The conservative South Carolina legislator raised the issue after viewing Roger Mappeltree's controversial exhibit of scientific photographs at the National Institutes of Health (NIH). The exhibit, "Various Imaging Techniques as a Means of Scientific Expressionism," is a collection of 50 photographs compiled by Mappeltree during his tenure as archivist at the National Medical Library on the main Bethesda, Maryland, campus of the NIH.

"I can't define science," Senator Holmes told his colleagues on the Senate floor, "but I know it when I see it, and this exhibit isn't science."

The photographs in the exhibit were selected by peer review from among 1,500 submitted by U.S. scientists. The nationwide competition was open to all researchers receiving Federal support. Senator Holmes says four of the photographs (reproduced here) are particularly offensive, calling them "porn masquerading as science." "The most objectional one," said Holmes, "is the one that looks like thingamajiggies." He was referring to the "transmission electron micrograph of a double-stage replica of two closely spaced dislocation pairs in silicon which has been chemically etched." Senator Holmes said that perhaps science organizations receiving Federal funding were "too well-endowed."

Quick Approval in Senate

The Senate has approved the measure by a vote of 99–1. (There was little debate, as Senators expressed concern that an extended investigation would leave members with insufficient time to attend their daily meetings with lobbyists.) Senator Aldo Matta of New Jersey registered the lone negative vote. "These pictures may very well be indecent," he told a television interviewer, "but science should be as free of values as the U.S. Senate."

President Is Concerned

If adopted, the Holmes bill would create an independent five-member panel of Senator Holmes's choosing, which would determine if Federally funded scientific work violated community standards for decency and obscenity. If the panel decided a project was without redeeming scientific value, grants would be cut immediately and all copies of the research, even those electronically preserved, would be destroyed.

The President is expected to sign the bill if certain national security issues can be addressed. The President was said to be distressed by ambiguity in the legislation's language whereby the entire Defense Department budget could be defined as "obscene."

Mappeltree's Provocative History

Roger Mappeltree, the 43-year-old archivist at the NIH's National Medical Laboratory, is no stranger to controversy. Five years ago his staff and budget were cut in half when he featured the work of Nobel Laureate Olga Hornsbee on the cover of the NIH's *Annual Re-*

view of Science. The photograph, *"in vitro* uriniferous modality with messianic substrate," has been described by Senator Holmes as "shocking scientific schlock." He described Hornsbee as "a deviant Lab Coat Lucy who seeks public funds to finance her porn."

Mappeltree himself is declining to speak with the press. He currently is archiving the millions of images that have been produced by U.S. reconnaissance satellites since 1964. Senator Holmes is now attempting to have the reconnaissance images seized and destroyed.

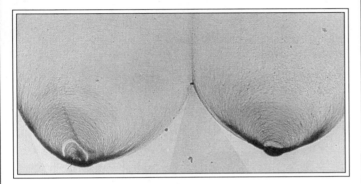

Demonstrators and Hubbub

As Senator Holmes was addressing the nearly empty Senate chamber, scientists critical of the antiporn measure marched outside the Capitol building. "Who needs all this tsimmes?" said one Hush-Puppy-clad microbiologist. "We have enough problems with the big-versus-little science fight."

A number of protesters held a "teach-in" for Washington schoolchildren, attempting to explain the controversy in terms the young people would understand. "We have mounted a multidisciplinary approach to studying the socioeconomic impact the bill will have relative to a cross section of the national populace," the children were told.

A counterdemonstrator sympathetic to Holmes called the Mappeltree pictures "erotica scientifica."

Ban Sought in Boston

The Mappeltree exhibition, which includes a daguerrotype of the first known suture and also includes the much-heralded work "Subliminal Messages in Gels," was scheduled to appear at the National Medical Library until the end of the summer. Now, however, that schedule is very much in doubt, as is the 12-city tour that was planned to follow it.

In Boston, the scheduled second stop on the tour, a spokeswoman for the Institute of Contemporary Art said her organization "makes it a practice never to get involved in anything controversial." She declined to say whether the Institute would go ahead with the exhibit.

Possible Effects on Education

In calling for Senate support for his "antiporn in science" bill, Senator Holmes is saying that if scientists are allowed "to use Federal money to make their filthy pictures, it could seriously ruin science education in this country. American science students will switch to soft, sissy sciences like anthropology." The Senator, a longstanding critic of the social sciences, suffered a political setback last year with a failed attempt at banning *National Geographic* from schools, libraries, and dentists' offices.

Responding to charges that his bill would violate the civil liberties of the nation's researchers, Holmes said, "My staff tells me that scientists aren't even mentioned in the Constitution."

Left to right: **Figure 1.** Scanning electron micrograph of the sporozoan parasite, M. cyrcumcystis. (Note that the geometric isomerism of this organism may be demonstrated by rotating the photograph 180°.) Photo: Merrie J. Mendenhall, University of Texas Southwestern Medical Center, Dallas. **Figure 2.** A transmission electron micrograph of a double-stage replica of two closely spaced dislocation pairs in silicon which has been chemically etched. The photomicrograph was made by Dr. George T.T. Sheng of Bell Laboratories, Murray Hill, New Jersey. It was formerly in the collection of David B. and Barbara T. Eisendrath of Brooklyn, New York. **Figure 3.** Two plant protoplasts caught in flagrante. Photo: David G. Davis, USDA Agricultural Research Service, Metabolism and Radiation Research Laboratory, Fargo, North Dakota. **Figure 4.** Magnetometers over San Francisco. The ring is a superconducting magnetometer; the bubble at the top is a liquid helium reservoir. The cylinder is a conventional flux-gate magnetometer. The entire apparatus was hung from a helicopter and flown over San Francisco Bay, circa 1980. Photo: Robert M. Rose of MIT.

ARCHITECTURAL SQUARING

Blair Charles
Raleigh, North Carolina

A proposed solution to certain architectural inefficiencies imposed by bias-block utilization of land holdings in urban areas is presented.

Illustrated below is the architectural squaring concept developed for the Title 21 Federal Urban Restructuring Program. The drawing presents views of the structure before and after squaring. The actual buildings depicted here are located in Cambridge, Massachusetts. The project was initiated under the auspices of a grant from the Reclamation Bureau's Concrete Utilization Authority, Eastern Regional Office, under provisions of Title 21, Section 912, Subsection D-90, Paragraphs 334, 335, 336, 337 and 339, under the August 1989 interpretation, as amended March 3, 1990.

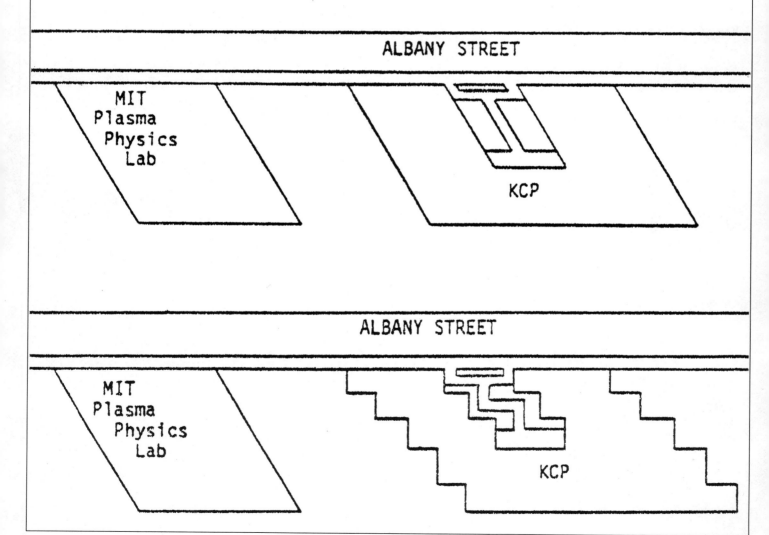

UTILITY OF MAGNETIC RESONANCE IMAGING IN PHILATELY

M. Hardjasudarma
Louisiana State University, Shreveport, Louisiana

Although uncommon, the use of diagnostic radiographic equipment outside the field of radiology is on record. Discoveries of fraudulent alterations or the presence of hitherto unrecognized masterpieces underneath more recent coats of colored substances in the field of paintings are prime examples.

We now present findings and results of recent work in which philatelic (Latin: *phil*—to like, to be fond of; *ateles*—salivating on gum. *Philately:* gum licking, or the hobby of stamp collecting) items were examined by magnetic resonance imaging.

The "British Guyana 1¢ magenta"

The world's most expensive postage stamp, still in its original locket, was subjected to standard imaging sequences. A confusing picture was produced, consisting of not one but four identical but seemingly randomly superimposed images. Besides the expected normal image, two were mirror image, and one image was both inverted and mirrored. We thought this represented a complex variant of the "wraparound" artifact, that is, the "wrap inside" variation. There remained no alternative but to remove the stamp from its locket to subject it to direct visual inspection. This initially met with considerable resistance on the part of its owner, who argued that breaking the seal on the locket would void any warranties given by the British Guyana post office. We succeeded in getting his consent when, upon calling the toll-free 800 number on the back of the locket, we found out that the phone number now belongs to a body shop in Demerara, Guyana.

The "stamp" as it turned out, was actually comprised of four specimens, folded as in Figure 1. In close co-

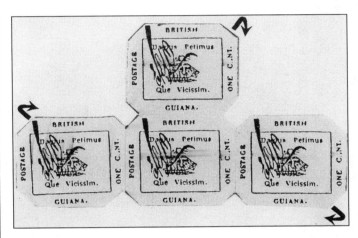

Figure 1. The "unique" British-Guyana stamp in an irregular block of four, folded according to directions indicated by arrows. Note perforations in the corners of each stamp.

operation with the Museum of Anthropology in Georgetown, Guyana, we were able to determine that the so-called "cut-corners" of these stamps were in fact produced by hollowed, square-pointed arrows used by the now-extinct Skwareperf tribe (Bow, 1990). The perforations measured 0.8:0.5 cm.

We postulate that the original seller, fearing that he would get less if it became known there were in fact four of these "unique" stamps, but loathe to separate them, folded three of them behind the fourth.

Incidentally, at least one of these perforating devices found its way to Saskatchewan, Canada, where it was later used to invalidate legal fee revenue stamps.

Gum Preservation

Philatelists are faced with the excruciating dilemma of having to preserve original gum in order that the stamp keep its value, while knowing full well that the gum will ultimately destroy the stamp (Lu, 1938). To discover

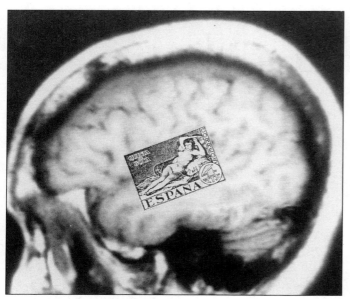

Figure 2. Goya's "Reclining Maja" tattooed on the operculum of the left cerebral hemisphere. Parasagittal T1-weighted image (TR 600/TE 20/2 Nex) of the brain.

whether we could separate the gum intact from the paper of the stamp, so that the two could then be stored individually, we put one VF-full OG postage stamp of Spain's Goya set into our MRI scanner.

We found that gum separation did not ensue, but we were left with a gummed piece of paper without a picture. This phenomenon remained unexplained until the next morning, when our first patient came to the MRI section of the radiology department for a brain scan. The note on the patient's requisition stated that there had been a recent change in behavior, namely depression, and ruled out organic cause.

Except for a reclining figure superimposed on his left operculum as seen on one sagittal T1W image (Figure 2), the examination was entirely normal. We have since conducted an analysis of the ink used to print the stamp and found that it had a high concentration of an iron pigment. This ferromagnetic property caused it to be lifted from the stamp and to be attached to the magnet during the initial scanning. Then, when our patient was scanned, the stamp image was transferred transcranially onto his left cerebral cortex, orthogonal to Penfield's homunculus.

We inspected the patient's scalp over the area of interest and found it to be visually intact. As a precau-

tion, he was put on broad-spectrum antibiotics for 2 weeks. A 6-month follow-up has seen our patient improve significantly, and we postulate that at least some of the iron particles acted like microneedles, in effect acupuncturing vital neurons and creating unexpected but beneficial results (Mackie, unpublished).

Acknowledgments

This manuscript is dedicated to my friends, Dr. K.W. Mackie, who taught me that humor need not be cerebral, and Dr. W.J. Maloney, who convincingly proved that cerebral humor need not be intelligent, and to so many of my fellow radiologists who doggedly insist on blurring the distinction between reality and hilarity.

REFERENCES

Bow EL. Adverse aerodynamic effects on the square-tipped arrow as a cause of starvation (doctoral dissertation). *Am J Point Counterpoint* 1990; 1:1.

Lu ES. Gummas: Adhesive and non-adhesive. *Internat Arch Vener Dis* 1938; 5473:17–71.

Mackie KW. *Aberrant and Unusual Pathways of Cerebral Stimulation.* [Unpublished autobiography].

CREATION OF THE UNIVERSE

This image appeared on television screens throughout North America during early 1991. The phenomenon is not well understood. Photo: Dr. Robert Richard Smith/Minimum Wage Art.

Styles, trends, and tidbits culled from leading research journals

by Alice Shirell Kaswell

Max Factor has developed a remarkable hygienic pump makeup that is hypoallergenic and undetectable. The pump provides coverage that is full of dimension. It is based on a weightless formula that helps to duplicate the natural tone of the skin with the precision of up to 300% finer color pigment dots per inch (DPI). Max Factor reports that observers conclude that it is radiant and perfect. Neutrogena reports that its researchers have stumbled on a cream that is noncomedogenic and that is in perfect balance with the lightest lipids. Chanel Research Laboratories has developed a remarkable formula for a complex of proteins and plant extracts. It is fortified. Sorbie has put confidence in a bottle.

Skin Exfoliation

Clarins has produced an exclusive formula that uses time-proven botanicals. Day after day, results become more visible. Clarins also reports an *eau dynamisante* fragrance and a *multi-senseur buste*/bust firming gel, for which it is necessary to always exfoliate the skin first.

Microspheres Contain Elements

Lancôme (Paris) enters the world of chrono-cosmetology. Lancôme's noctosome system *renovateur de nuit* provides for the accumulation of a precious nightly reservoir of niosome microspheres containing specific elements. The unique time-released niosome system technology allows accurate, targeted, and gradual transport of its special ingredients in time-released fashion to surface layers of the skin. (Lancôme also reports yet another breakthrough in microbubble technology.)

Behavior Pattern of Durability

Clinique has perfected a pigment that will not change colors and that can be combined with lubricants that do not blur. The resulting substance has a behavior pattern of durability with no stain or dryness drawbacks. The laboratory also reports that the substance has long-lasting table manners and says that this has been verified with a computer.

Time Recharging and Complex

Estée Lauder has obtained interesting results with a time zone moisture recharging complex. Among other properties it is able to control irrigation on a daily basis. Lauder is also making use of a microemulsion formula. Elizabeth Arden has developed a ceramide time complex capsule based on its pure potent ceramide 1 formula.

Nagging Has Failed

The research journal **Cosmopolitan** reports (vol. 205, no. 5, p. 84) that Antabuse, selenium, and certain nitroglycerine pills give breath a characteristic odor. The investigator, Susan Okie, M.D., suggests a line of research based on brushing the tongue. The same journal contains the finding of another investigator (Irma Kurtz) that nagging has failed. Kurtz also presents preliminary results concerning a woman whose home is a mess and who is a hypochondriac. A report on p. 208 recommends that people who stand on their feet should obtain a shock absorber. Investigator Peggy Nichol reports (pp. A10–12) on the efficacy of enhancing a deep-penetrating conditioner by adding caviar. Nichol also reports that petroleum jelly prevents cornmeal from scratching.

Time vs Torque

A biomechanics research paper appearing on p. 32 of the June 1989 issue of the research journal **New Woman** reports that the longer you leave—not the tighter you wind—the rollers in your hair, the longer your curls will last.

Findings on Fluids

Kathryn Keller's seminal research on water (H_2O) is written up (p. 113 ff.) in vol. CLXXV, no. 1 of the research journal **Redbook**. The report's most salient finding is that H_2O ferries nutrients and oxygen to cells via the bloodstream and lymphatic system. Test subjects yielded impressive results by urinating every 3 or 4 hours.

Smart Soup, Random Wigs

The December 1990–January 1991 issue of **Modern Maturity** (vol. 33, no. 6) describes (p. 69) an artificial intelligence breakthrough in which Campbell's scientists have created eight soups that understand what you want and what you don't want. The issue contains two other brief but cogent reports, both on p. 93. The first describes a computational algorithm that turns your life experiences into a legacy of love and treasured wisdom. In the second report, investigator Paula Young presents a research protocol under which each month she randomly selects 100 subjects and asks each to choose a natural-looking fashion wig.

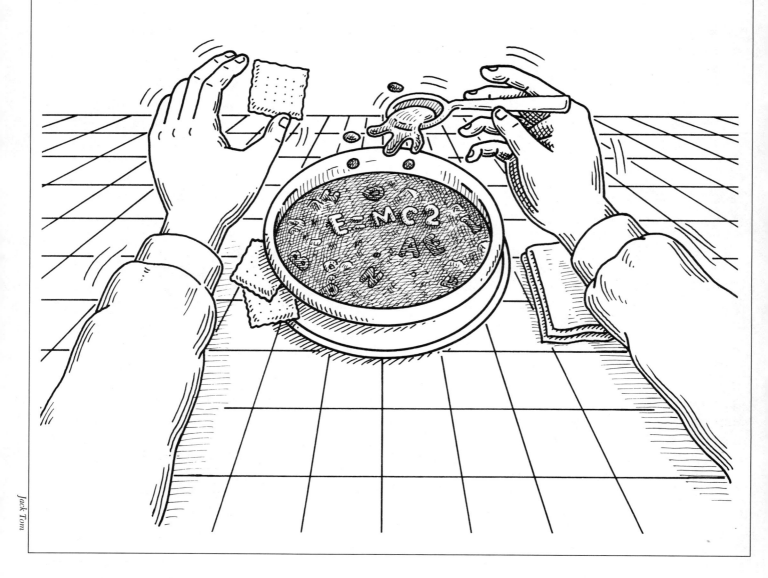

Jack Tom

CHAPTER 3

FOOD AND ITS HABITS

A DECADE OF BURNT FOOD

Rachel Ehlo Whirster
Somerville Massachusetts
Photographs by Roland Sharrillo

Our understanding of burnt food has increased tremendously in the past decade. Much of this progress is due to research conducted at the Museum of Burnt Food (Figure 1).

The exhibit wings of the museum house more than 49,000 scorched, singed, seared, and charred items of staggering diversity, ranging from carbonized poultry and soy dogs (Figure 2), burnt venison, pork, fish, and cider (Figure 3), to the celebrated thrice-baked potatoes (Figure 4).

The Hall of Burnt Toast alone contains more than 2,000 specimens (Figure 5). The Rice and Wheat Gallery also boasts an impressive number of items, many of which are as yet uncatalogued. One wing of the museum is devoted solely to burnt legumes. A newly renovated Hall of Charred Condiments is scheduled to open next May.

The research facilities occupy their own buildings and are not normally open to public viewing.

The museum's research staff has published more than 90 books on various aspects of burnt food and is credited with authorship of more than 850 papers that have appeared in research journals since 1982. Several books for non-specialists, including *The Joy of Burnt Food, The Carbon Diet,* and *Hyperorganic Cooking,* have aroused the public's interest in postnutrition science.

The museum was founded by Dr. Deborah Henson-Conant, a science historian who never lost her childhood

Figure 2. Coal-fired soy dog, circa 1989.

fascination with organic chemistry. A 1981 accident involving apple cider (Figure 3) inflamed in her a passionate curiosity about burnt food.

The museum, located in Somerville, Massachusetts, is visited annually by more than 25,000 people. The research staff includes four postnutrition food scholars and an extensive support staff. The exhibit wings and restaurant pavilion employ seven full-time educators and food service workers.

Figure 1. Dr. Deborah Henson-Conant, founder and curator of the Museum of Burnt Food. Henson-Conant is shown leading a recent seminar on the numerical analysis of incinerated cheeses.

Figure 3. Burnt apple cider, 1981. This is the specimen that inspired the museum's founding.

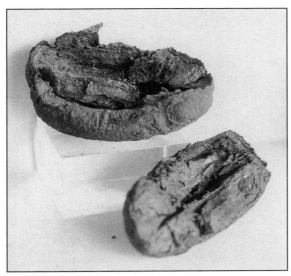

Figure 4. Thrice-baked potatoes, circa 1984.

Figure 5. An exhibit from the Old Toast Collection. The item on the right is a burnt toast specimen discovered in Glasgow, Scotland, in 1983. Carbon dating indicates that it was formed in the late 20th century. The item on the left is an artist's reconstruction of the toast as it may have appeared prior to incineration.

JIR RECOMMENDS

Articles, books, and other communications that merit your attention

Compiled by Stephen Drew, Norman D. Stevens, and X. Perry Mental

"Bread-Making as a Source of Vaginal Infection with *Saccharomyces cerevisiae*: Report of a Case in a Woman and Apparent Transmission to Her Partner," by J.D. Wilson, B.M. Jones, and G. R. Kinghorn, in *Journal of Sexually Transmitted Diseases*, Jan–Mar 1988, 35–36. (*JIR* thanks Mark Hochhauser for bringing this to our attention.)

"Chicken soup rebound and relapse of pneumonia: report of a case," by N.L. Caroline and H. Schwartz, *Chest* 1975; 67:215–216.

Dun Dun noodles at Mary Chung's Restaurant, Central Square, Cambridge, Massachusetts. The official food of the *JIR* editorial staff, the noodles are irreproducible in the finest sense: No one else has been able to reproduce the characteristically addictive flavor. Mary Chung's is an important meeting site for scientists and engineers in the eastern United States.

"The effect of away-from-home eating on the consumption of fluid milk." Published by The Department of Agricultural Economics and Rural Sociology, The Pennsylvania State University, University Park, Pennsylvania, November 1990.

"Effects of drinking hot water, cold water, and chicken soup on nasal mucus velocity and nasal airflow resistance," by K. Sakethoo, B.S. Januszkiewicz, and M.A. Sackner, *Chest* 1978; 74:4. (*JIR* thanks Bob Hoffman for this.)

"*Memoir of a Thinking Radish*," by P. Medawar, Oxford, England: Oxford University Press, 1986. (*JIR* thanks Gary D. Miller.)

"Why do vegetarian restaurants serve hamburgers? Toward an understanding of a cuisine," by L. Guion-Rosenberg, *Semiotica* 1990; 80-½: 61–79. (*JIR* thanks Laurie Rothstein and Jacqueline Baum for bringing this to our attention.)

SURVIVAL STRATEGIES AMONG ANIMAL CRACKERS

Nabisco Brands, Inc.
Attn: Manager, Customer Service
Barnum Animals
East Hanover, NJ 07936

Dear Sir or Madam:

We bought a box of Barnum's Animals crackers today and would like to make several comments, which we hope you will accept in the spirit of constructive criticism:

1. We could not help noticing that many of the crackers were in several pieces. While we did not make a detailed survey, it appeared that this was particularly true of the prey. Also, a few of the predators seemed larger than others of their kind. We think the problem here is that you have packaged the predators and prey in the same small box. A simple textbook on ecology will explain why this results in fragmented prey.

2. You indicate on the bottom of the box, "When writing to us, please enclose the top of the package." We are not doing this, however, because the top of the package has an attractive folding clown on it. We spent quite a while rummaging through the animal cracker boxes at the A&P, trying to decide on whether we wanted a clown, ringmaster, etc. The clown was clearly superior, in our judgment, and we want to keep it. Furthermore, we feel that this clown has been through enough as it is, for he suffered greatly when we tried to open the package according to your instructions, "To open, pinch here."

As we said, we don't want you to think we are just being negative, so here are some positive comments and suggestions for improvements:

1. You are to be congratulated for having the clown on the outside of the box, and not on the inside, since there's no telling what the predators would do to the clown, given what they've done to the prey.

2. Several possible solutions to the predator-prey problem come to mind. You might have a larger box, perhaps with compartments. Also, do you feed the predators just before packing? Another possibility is to have several different types of products, such as "Barnum's Prey," "Barnum's Predators," "Barnum's Clowns," "Barnum's Don't Cares," etc.

Thank you for your interest and concern on these matters.

Sincerely yours,
Mr. and Mrs. Robert L.
Feldman and family
(David and Elana)
Ithaca, New York

MULTIVARIATE ANALYSIS OF SPAGHETTI SAUCE STAINS

J. Fairbanks Dow
New Port Richey, Florida

Abstract
The analysis of the amount of spaghetti sauce stains on shirts of different colors and patterns reveals an interesting correlation.

1. Introduction
My colleagues and I have been interested for years in an interesting phenomenon: Is it true that spaghetti sauce splatters tend to cluster more readily on fabric lacking colors or patterns?

2. Methodology
2.1. Survey
2.1.1. Data Gathering
2.1.1.1. Restaurants
2.1.1.1.1. People
2.1.1.1.1.1. Eating
2.1.1.1.1.1.1. Spaghetti
2.1.2. Selection

Five Italian restaurants[1] were selected at random in five randomly determined cities[2] in a randomly selected country in North America.[3] Survey technicians were dispatched to each of these cities, and on a random evening[4] they entered the Italian restaurants.

2.2. Survey Hardware
Using handheld infrared detection devices, the survey team scanned patrons as they left. Special emphasis was given to the area between the collar and the waist on both men's and women's clothing. Patrons with jackets or wraps were requested to open them.

3. Analysis
Figures 1 and 2 show the results of the survey. Controlling for both sex and location within the randomly selected North American country, it is very apparent that our data support the original hypothesis.

Figure 1. Controlling for sex.

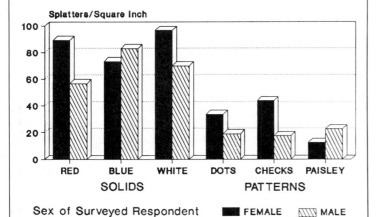

Figure 2. Controlling for location.

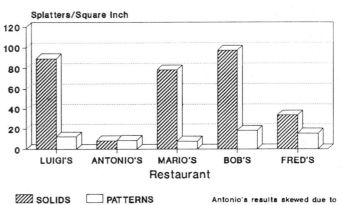

4. Conclusion

There is a definite correlation not only between the color of a shirt and the number of splatters per inch, but also between the pattern (or lack thereof) and the amount of splattering.

5. Future Research

Given the results of our original efforts, my colleagues and I have proposed further study along the following lines: Does the fabric of the shirt have anything to do with the splattering patterns? Would rayon differ from Dacron, for example, or would Dacron differ from cotton? Is there a statistically supportable causal relationship within the splattering/fabric pattern correlation? Are there "sauceones," subatomic particles within white shirts that attract tomato sauce? What if we connect the dots within the splattering patterns? Can the infrared device be removed from the face of the Cincinnati technician who used the phrase "Spread 'em" to get a patron to open his sport coat?

1. Luigi's, Antonio's, Mario's, Bob's, and Fred's.
2. Los Angeles, Cincinnati, New York, New Orleans, and San Francisco, respectively.
3. The United States.
4. Thursday.

NOBEL THOUGHTS

Profound insights of the laureates

Marc Abrahams

Fred Robbins is University Professor Emeritus and Dean Emeritus at Case Western Reserve University Medical School. In 1954 he was awarded the Nobel Prize in Physiology or Medicine for his work in developing methods for growing viruses in cultures of various types of body cells. This work led to the later development of many vaccines. He spoke, via telephone, from his office in Cleveland.

What is your favorite food for lunch?

Soup. I don't like creamy soups. I don't like watery soups. Anything in between.

Do you recommend that people take a lunch from home or buy it at the lab?

I believe in the old Rockefeller University policy. They eat in a dining room where they can have some good conversation. I've been accused of being a table-hopper.

Do you have strong views about dessert?

Yes. I *like* it.

Do you have any advice for young people who are entering the field?

I want to say, "Stay out of it," but I don't think that's a very good idea. No, the best advice I can give is to find the right guy to work with. Make some inquiries. Look for a person who gives some time to the people who work with him. It should be a person who's concerned with the welfare of those who work with him, and a person whose research is thought superior.

A REFINED ECONOMETRIC ANALYSIS OF WINE

Picrocole Rashcalf

Pullman, Washington

In January 1990 an issue of the Sunday **New York Times** carried a front page article on an economist who estimated the economic value of French Bordeaux via a multiple regression model rather than by sniffing it, swilling it around on his tongue, rinsing his teeth, rolling his eyeballs, and expectorating.

Basically, this economist says that the esthetic, and therefore the economic, value of, say, a Lafite-Rothschild (Pauillac) 1841 (selling at auction in December 1988 for U.S. $24,000 a case) is a function of dirt, water, and the weather. His claim is that about 80% of the market value of the red Bordeaux and other red wines that enter speculative trade is due to these variables. Unscientific. Ridiculous. Sacrilegious.

There are fundamental reasons to justify the highly personalized hatred and contemptuous disdain that have been heaped on this particular economist's head in subsequent letters to the **New York Times**.[1] Simply put, he totally ignored the ineffable qualities that make up a great wine. The correct categorical variables could have transformed a mere exercise into a straightforward scientific analysis.[2]

He need only have read any decent book on wine connoisseuring to discover the appropriate descriptors. Below is a short list discovered through only a few moments' examination of such a volume.[3]

Complex and interesting = 1; 0, otherwise.
Generous = 1; 0, otherwise.
Soft = 1; 0, otherwise.
Tastes like socks = 1; 0, otherwise.
Austere = 1; 0, otherwise.
Clean = 1; 0, otherwise.
Robust, assertive, and immediately obvious = 1; 0, otherwise.

Common as dirt = 1; 0, otherwise.
Thick = 1; 0, otherwise.
Quiet and newly born = 1; 0, otherwise.
Great = 1; 0, otherwise.
Gracious = 1; 0, otherwise.
Spicy sweetness that is almost oriental = 1; 0, otherwise.
Elegant = 1; 0, otherwise.
Big = 1; 0, otherwise.
Has just a soupçon of the toe jam of a French peasant = 1; 0, otherwise.
Clumsy = 1; 0, otherwise.[4]
Rotgut = 1; 0, otherwise.
Possesses breeding and fullness of body and power = 1; 0, otherwise.
Randy as a goat = 1; 0, otherwise.
Coarse and potent = 1; 0, otherwise.
Has distinctive barnyard odors = 1; 0, otherwise.[5]
Tastes like cat urine = 1; 0, otherwise.

These concepts have a richness of meaning that transcends water, dirt, and the weather.[6]

---

1. See the **New York Times**, Letters to the Editor, March 18, 1990.

2. An hedonic value function, as it were.

3. This is only a partial list taken from my wine encyclopedia *Wines of the World*, Andre L. Simon, editor, as well as some descriptors gleaned from the popular press.

4. In all fairness to our economist, just what is a clumsy wine? I, Picrocole Rashcalf, have never seen a clumsy wine. Do the bottles filled with such wine spontaneously fall off of the wine rack? If they do, how does one taste them, unless by licking up the wine from the floor. And why would a sane person want to do that only to verify that the wine was clumsy when the very act of the bottle falling off the rack has demonstrated the characteristic? This is perhaps too mystical. Should one omit this variable?

5. According to **Money** magazine, this and the following variable are concepts developed by a well-respected American wine connoisseur.

6. The economist missed the essence of the problem. He could have formed interaction terms among these variables with the usual multiplication of two or more regressors. Consider the combination of "robust, assertive, and immediately obvious rotgut" or "elegant and randy as a clumsy goat" or "quiet and newly born, yet with the breeding and fullness of body and power" or "tastes like thick, clean socks." Then, too, of course, there is "coarse, potent, and randy as a clumsy goat" and "tastes like cat urine with a spicy sweetness that is almost oriental."

DISTRIBUTION OF DENNY'S RESTAURANTS IN THE CONTINENTAL UNITED STATES

Jim Cser
Aloha, Oregon

Abstract
Meaningless statistics about the distribution of Denny's Family Restaurants are compiled and analyzed. Calculations of Denny's per capita and Denny's per unit area are made for each state. Important vacation tips are presented, but you'll have to read the study to find out what they are.

Introduction

Every year, millions of Americans jump in the car and go on vacation.[1] One of the most important parts of vacationing is finding somewhere to eat. The choice of restaurants can be bewildering at times, especially in unfamiliar territory. The decision must be made with great care, for many an otherwise happy vacation has been ruined by a burnt hamburger or a salad topped with unusual foreign objects.

Nothing is more reassuring than pulling into a Denny's Family Restaurant: You know exactly what the in-

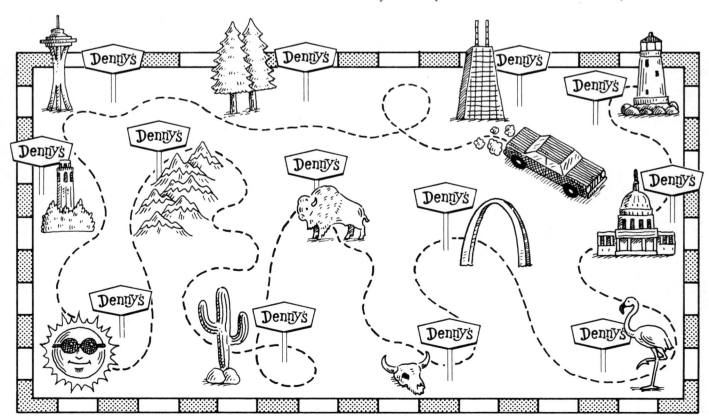

Jack Tom

Table 1. Denny's distribution by state.

STATE	AREA (SQ. MILES)	POPULATION (MILLIONS)	DENNY'S	DENSITY	CROWDING INDEX
AL	50,766	4.102	8	0.158	1.950
AR	53,187	2.395	5	0.094	2.088
AZ	113,510	3.489	46	0.405	13.184
CA	156,297	28.314	328	2.009	11.584
CO	103,598	3.301	30	0.290	9.088
CT	4,872	3.233	9	1.847	2.784
DE	1,933	0.660	3	1.552	4.545
FL	54,157	12.335	140	2.585	11.350
GA	58,060	6.342	21	0.362	3.311
IA	55,965	2.834	6	0.107	2.117
ID	82,413	1.003	5	0.061	4.985
IL	55,646	11.614	54	0.970	4.650
IN	35,936	5.556	24	0.668	4.320
KS	81,783	2.495	10	0.122	4.008
KY	39,674	3.727	4	0.101	1.073
LA	44,520	4.408	10	0.225	2.269
MA	7,826	5.889	10	1.278	1.698
MD	9,838	4.622	24	2.440	5.193
ME	30,995	1.205	3	0.097	2.490
MI	56,959	9.240	41	0.720	4.437
MN	79,548	4.307	14	0.176	3.251
MO	68,945	5.141	29	0.421	5.641
MS	47,243	2.620	2	0.042	0.763
MT	145,338	0.805	0	0	0
NB	76,639	1.602	6	0.078	3.745
NC	48,843	6.489	4	0.082	0.616
ND	69,249	0.667	1	0.014	1.499
NH	8,922	1.085	2	0.224	1.843
NJ	7,468	7.721	17	2.276	2.202
NM	121,336	1.507	12	0.099	7.963
NV	109,895	1.054	11	0.100	10.436
NY	47,379	17.909	31	0.654	1.731
OH	41,004	10.885	50	1.219	4.606
OK	68,656	3.242	13	0.189	4.010
OR	96,187	2.767	17	0.177	6.144
PA	44,892	12.001	34	0.757	2.833
RI	1,054	0.993	0	0	0
SC	30,207	3.470	8	0.265	2.305
SD	75,956	0.713	0	0	0
TN	41,154	4.895	11	0.267	2.247
TX	262,015	16.841	90	0.343	5.344
UT	82,076	1.690	11	0.134	6.509
VA	39,700	6.015	17	0.428	2.826
VT	9,273	0.557	2	0.216	3.591
WA	66,512	4.648	68	1.022	14.630
WS	54,424	4.855	16	0.294	3.296
WV	24,124	1.876	1	0.042	0.533
WY	96,988	0.479	4	0.041	2.088

terior looks like, you know exactly what's on the menu, you know exactly what the coffee is going to taste like, and, of course, you know that they never close.[2] Therefore, when I am on the road in some obscure part of the country, the question foremost on my mind is, "where's the nearest Denny's?" Having obtained a copy of the master list of all the Denny's in the world,[3] the answer to this question, at least, is always at my fingertips.

However, while leafing through this fascinating booklet, a couple of other questions came immediately to mind. First, what states should I visit, in order to have the maximum probability of finding a Denny's? If I am going to plan a proper vacation, I am going to make sure that there will always be a Denny's nearby when I need one. Second, which states have the most Denny's per capita? Even if I do find a Denny's, there is little

point if I am forced to wait around all day to get served. In this study I hope to resolve these burning questions, to make everyone's vacation a little bit better, and, being a graduate student, to appear as if I am working on my thesis.

For this study, I am restricting myself to Denny's locations in the continental United States.

Table 1 shows the area,[4] population,[4] number of Denny's,[3] Denny's density (Denny's/1,000 square miles), and Crowding Index (Denny's/1,000,000 people) for each state. In terms of sheer numbers, California is the clear winner, with a whopping 328 Denny's. Visiting all of them on a single vacation presents quite a temptation, and will be the subject of future research.

Surprisingly, three states (Montana, South Dakota, and Rhode Island) have no Denny's whatsoever. In light of the concerns expressed above, I would strongly recommend against setting foot in any of these states. A possible exception might be South Dakota, home of Mount Rushmore, Wall Drug, and the annual Harley-Davidson motorcycle convention.

Florida has the greatest Denny's density, with 2.585 Denny's/1,000 square miles. This translates to a Denny's mean free path of 19.67 miles, or 0.065 tanks. (One tank = 300 miles; mileage may vary.) Of the states with Denny's, North Dakota has the lowest Denny's density, with 0.014 Denny's/1,000 square miles, giving a mean free path of 267 miles, or 0.89 tanks. As this distance is greater than 0.5 tanks, a round trip will certainly leave you stranded somewhere, which is generally to be avoided.

The least crowded state is Washington, with 14.63 Denny's/1,000,000 people, or one Denny's for every 68,353 people. This many people sharing one restaurant might sound like a lot, but if you have ever tried to go out to breakfast on a Sunday morning, you will agree that this number is correct to within an order of magnitude. The most crowded state is West Virginia, with 0.533 Denny's/1,000,000 people, or one Denny's for every 1.88 million people, a situation no restaurant patron should have to endure.

Finally, the degree of vacation desirability is determined by the Density-Crowding Product. The winner is Florida, which just edges out California, 29.34 to 24.32, respectively. Of those states with Denny's, the lowest Density-Crowding Product goes to North Dakota, just edging out West Virginia, 0.021 to 0.022.

Conclusion

In order to have a vacation with an acceptable number of uncrowded Denny's restaurants along the way, the best states to visit are Florida, California, and Washington. Traveler's advisories go out for Montana, Rhode Island, North and South Dakota, West Virginia, North Carolina, Missouri, and Wyoming. This research was supported by the American Automobile Association and the Oregon Institute of Recreation, Department of Vacation Science.

---

1. "Every year millions of Americans jump in the car and go on vacation," *USA Today,* June 12, 1985.

2. Actually, they were closed for Christmas in 1988, which required them to install previously unneeded locks in the front doors, at a cost of roughly $4,000,000.

3. *1990 Denny's Travel Guide,* Denny's Press, Minneapolis, MN.

4. *1990 World Almanac and Book of Facts.*

APOLOGY TO OUR READERS

Because of an unfortunate error, the fractal potatoes advertised in *JIR* 36:2 are no longer available for purchase. Our 1991 fractal potato crop was accidentally shipped to a company that used them to produce powdered mashed potatoes. We are endeavoring to mail refunds to everyone who has placed an order. We apologize for any inconvenience and disappointment this may cause.

FRACTAL BREAKFAST CEREAL: A CHAOTIC DYNAMIC SYSTEM FORTIFIED WITH EIGHT ESSENTIAL VITAMINS AND MINERALS

J.S. Ottaviani
Ann Arbor, Michigan

Abstract

A summary of the results from an interdisciplinary study is presented, confirming the efficacy of applying fractal geometries to the classic breakfast cereal problems of saturation and packing.

Method

A prototype cereal, "Fractals," was developed and tested for its ability to solve two classic breakfast cereal problems. The first, saturation (sogginess), has received some attention in the industry. Rice Krispies®, Crispex®, and Team® are typical of early attempts to delay the onset of sogginess. Part 1 of this report gives results of a comparison of sogginess as a function of time between a typical cereal and "Fractals." The second problem, more of an engineering/marketing concern, has its roots in packaging theory. Part 2 reports what, if anything, is meant by that.

Part 1: Saturation

Figure 1 shows a comparison of time to saturation (inedible sogginess) between Brand "X" and "Fractals." The effects of a quasi-infinite surface area are dramatic. The experiment began at t_o, when 4 times the serving size of each cereal was combined with 6 servings of milk.[1] At $t_{f(x)}$, Brand "X" was at an unacceptable sogginess level[2] and had to be composted. "Fractals" showed little increase in sogginess well beyond this point. The ex-

periment was stopped at $t_{f(f)}$. "Fractals" were indeed soggy at $t_{f(f)}$, but by this time the milk has turned a complex color, exhibiting the rarely observed second-order Carrell effect.[3] The paucity of data points for "Fractals" is regrettable but unavoidable in a study of this type. Having been raised according to the instant-gratification-entertainment paradigm on which the experiment draws, the experimenter grew bored with taking data, turned on the TV, and ceased taking measurements. So $t_{f(f)}$ (and the curve leading to it) is the best-guess[4] value of when sogginess occurred.

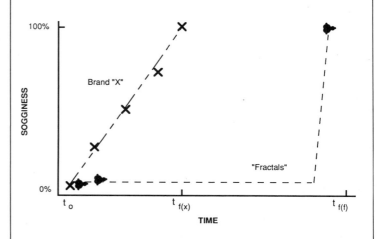

Figure 1. Comparison of a "typical" cereal and "Fractals."

Part 2: Settling

A phrase that commonly appears on cereal boxes is "This package is sold by weight not volume. Some settling may have occurred during shipping and handling." "Fractals," with their nonlinear dynamical roots, need never be packaged in boxes displaying this type of statement. "Fractals" never settle. This does not, however, result in unacceptable numbers of "topologic refugees."[5] This term refers to cereal that is found between the box and the lining. Time-lapse photography (Figure 2) shows that "Fractals" do not migrate through this lining via quantum tunnelling or any other anomalous effect.

Conclusions and Directions for Future Research

"Fractals" imply breakfast. "Fractals" sog slowly. "Fractals" turn the milk a complex color. "Fractals" do not settle down. "Fractals" stay where they belong.

Future research should address the following questions: How many bowls of "Fractals" does it take to equal one bowl of Total®? Can strange attractors be applied to advertising and marketing? How many proofs-of-purchase should be required for a free Julia set?

Figure 2. "Fractals": Non-migration over time.

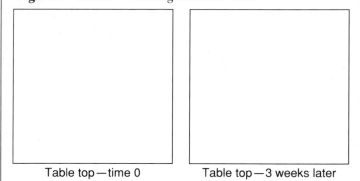

Table top—time 0   Table top—3 weeks later

1. The experimenter has never eaten just one serving/bowl.

2. For the mathematically inclined, the technical term is "gross."

3. Named for the pioneering cereal-fractalogist M.C. Carrell.

4. Again, for the mathematically inclined, the technical term is "fudged."

5. Dr. Science's answer to the query, "Occasionally when I open a box of Cheerios® I notice a few outsiders (sic) of the bag inside. How did they get there? Are they safe to eat?" in ***Orbit*** 1 (November 8, 1990), p. 23.

THE DARK SIDE OF COFFEE

Pilk John Hannay
Bethesda, Maryland

Our analysis at the National Agronomic Laboratory of the 1990 world coffee crops yielded results that are consistent with those of previous years' surveys,[1] confirming that, statistically, coffee has a dark side.

The 1990 survey indicated that a statistically significant percentage (in the aggregate, between 52 and 55%) of coffee beans exhibit a distinct asymmetry in the surface distribution of coloring. In plain language: one side of the bean is darker than the other side.

The finding that coffee has a dark side is true only of coffee of "good quality" and does not apply to coffee grown in climatically inappropriate regions. In those regions (such as Sweden), a distinctly different asymmetry holds: a slight (between 51 and 54%) statistical preponderance of coffee beans has a light side.

1. The year 1979 did not conform to the pattern, however. In that year, anomalous weather conditions combined with petroleum-price-related transportation disruptions to alter the coffee crop compositions available in the United States.

COOKING WITH POTENTIAL ENERGY

R.C. Gimmi and Gloria J. Browne

Tucson, Arizona

Introduction

The world's supply of combustible energy resources, such as wood and fossil fuels, is rapidly diminishing. If civilization is to continue, we must quickly develop alternative non-combustible energy sources that can be used for basic human needs. Heating, transportation, and cooking will be of paramount concern.

This new energy must be cheap, safe, and of such low technology that it can be easily understood by the layperson. The obvious solution to any future energy crises is the development and use of potential energy. This form of energy is ubiquitous in nature; it is renewable; and, to date, it has largely been untapped.

All elevated objects possess potential energy. When an object is dropped from a height, its potential energy is converted into kinetic energy. Upon impact with the ground, this kinetic energy is converted into heat. This phenomenon may be described with the energy balance

$$\Delta Z \frac{g}{ge} = Cp(\Delta T).$$

If the change in height (ΔZ) can be made sufficiently large, significant amounts of heat can be generated. Everyday tasks like cooking could be made safe, simple, and energy-efficient.

In our experiment, we proved that this can be done. In 6 hours, we partly cooked a 25-pound turkey with potential energy.

Materials and Methods

There was considerable debate among the investigators as to what we should cook. A goose was suggested but we finally settled on a defrosted 25-pound turkey.

At 9:00 A.M., an undergraduate carried the thawed bird up the stairs to the tenth floor of the University of Southern Arizona administration building. From this vantage point, he flung the bird from a ledge.

Immediately after the turkey impacted on the pave-

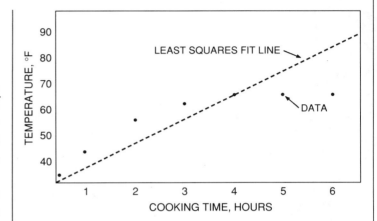

Figure 1. Rate of cooking for turkey. Initial turkey temperature, 32°F; Final turkey temperature, 65°F. Ambient temperature, 68°F.

ment, the investigators inserted a thermometer into the carcass and recorded the temperature. The assistant then ran down the stairs to retrieve the bird.

This process was repeated 72 times in 6 hours with the same turkey.

Results

Data are shown in Figure 1. A least squares fit method was used on the data. The rate of temperature increase for the turkey was 8°F per hour.

Discussion

At 3:00 P.M. the last of our funding[1] was consumed and the experiment was halted. By extrapolating from our existing data it was obvious that the turkey would have reached a temperature of 400° F in approximately 46 hours. While this is admittedly a "slow cook," our critics will have to concede that we did not burn any fuel.

The principal investigator sampled some of the partly cooked bird. He found it was "somewhat cool" but comparable to any of the cooking his spouse has done at home. He also reported that the meat was very tender.[2]

1. $11.07 for the turkey; $1.93 for bananas that were periodically given to our assistant.
2. Viscosity 94 centipose.

*Contains 100% gossip
from concentrate*

Compiled by Stephen Drew

Tracer of Lost Socks

It is now possible to reliably locate lost socks. New stockings are being manufactured with special VLSI microchips along the instep. Each sock has its own unique identifying code. Socks can be located, to an accuracy of within 2 m, by a method of passive communications with an orbiting satellite. The method does not work with old socks that are not equipped with the VLSI circuitry.

The SRLS (Satellite Retrieval of Lost Socks) system was developed by Josiah S. Carberry of Brown University. Carberry, who is best known for his pioneering work in the field of psychoceramics (i.e., cracked pots), has made salient contributions in many areas. Some observers are skeptical about the system, seeing it as an overly elaborate and expensive solution to a problem that, although thus far intractable, might someday yield to a less technically complex solution.

Power Coffee May Increase Osteoporosis Risk

Business executives' craze for "power coffee" may be putting their bones at risk. Power coffee is coffee that has been brewed in the usual way, and into which one (or sometimes two) teabags have been placed. A report that will be submitted to the **American Journal of Epidemiology** concludes that regular consumption of power coffee may increase drinkers' risk of osteoporosis, a condition involving bone loss and embrittlement. Osteoporosis contributes to the high rate of hip fracture among the elderly.

In some executive suites, a daily competition now takes place, with CEOs, CFOs, and senior vice presidents allowing their teabags to steep in coffee for long periods of time. Researchers examined data on 310 executive males, aged 52–65 years, and 42 executive females, aged 32–45, in the United States and France, and also made use of data from historical files in Norway, Japan, and Brazil (power coffee enjoyed a brief vogue in the latter three countries during the 1920s). Both sets

of data were also found to square with animal studies showing that power coffee increases urinary calcium excretion and may inhibit gastric absorption of calcium.

Power Coffee and the Berlin Wall

Another study, organized by a consortium of German universities, will examine the idea that power coffee increases the risk of heart disease. Researchers will survey 45,589 male dentists, optometrists, osteopathic physicians, pharmacists, podiatrists, and vegetarians, none of whom have a personal history of cardiovascular disease. Half of the volunteers live in the region that formerly was East Germany. Prior to November 1989, East German citizens had no access or exposure to power coffee.

Contraceptive Manufacturer Pulls "Happy" TV Ads

Bowing to anticipated pressure from religious leaders, the manufacturer of a popular brand of contraceptive jelly has instructed its advertising agency not to run a series of lighthearted television advertisements. The ads, which have been scheduled for broadcast on American television, featured the Bobby McFerrin song "Don't Worry, Be Happy."

Strategic Motion Initiative

The United States government will undertake a massive research effort to develop perpetual motion technology, if a leading physicist succeeds in his efforts to persuade key Defense officials. The physicist, who has played a vital role in the development of emerging defense technologies in the US since the 1940s, has enlisted the active support of a former U.S. President and several current congressional leaders. A recently retired British Prime Minister is also expected to join the effort.

Subliminal Comfort

Stiff arms and legs, poor blood circulation, couch sores, stale air. For these and other reasons, waiting rooms often wear out the people they were meant to welcome. That may soon change. Keisuke Terezawa of Tokyo's Design Functionality Corporation has discovered an array of techniques for preventing the tedious effects that people dread. In his test lounge, Terezawa uses couches that gently, semi-randomly vibrate; light levels that shift subtly at irregular intervals (the color composition of the light is also varied); varying subaudible vibrations in the air; varying ionization and scent manipulations in the air-conditioning system; and other methods to subtly enliven the plight of the waiting guest. The keys, says Terezawa, are subtlety and variability. The various changes in the tactile, visual, olfactory, and electrical environment are all performed in slight in-

Jack Tom

crements and at irregular intervals. Terezawa found that changes which are noticeable or periodic are perceived as disruptive rather than comforting. Devices incorporating his principles are expected to appear in corporate waiting rooms within 3 years.

New Standard for OLP Formula

The longstanding disagreement concerning the substance OLP has been settled by the United Nations Organization for Chemical Formulaic Standards. OLP is a pheromone produced by elderly mammals. OLP's ancient folkname, from which the formal name is derived, is "old laddie powder" or "old lady powder." OLP was first artificially synthesized in 1966 by F. Boisse Laboratories of Montpelier, France. Synthetic OLP has become a key component in the manufacture of many products, including magnetic recording tape.

The formula has been in dispute since 1972, when Boisse Laboratories lost control of the patent to a consortium of British, Italian, and Romanian interests. Litigation involving at least 17 countries, complicated by an unrelated series of administrative disputes among several international standards organizations, led to the *de facto* acceptance of several competing standards for defining the OLP formula. The new standard is expected to have an effect on OLP prices, lowering them or possibly raising them.

Growing Chicken Little

Nicholas Natalee, whose pioneering studies of growth have spanned such varied fields as astronomy, human

---

SLEEP RESEARCH UPDATE

- AG has stopped sleeping with DF and is now sleeping with DL.
- DF now sleeps with FP and LB, but not with GS except when the temperature drops below 43° (Fahrenheit).
- NVonC is now sleeping with BB.
- JS is now drinking hot milk with onions before going to bed. He is not sleeping with anyone.
- VM is sleeping with Dr. SL, Dr. BR, Dr. IB, Dr. LH, and Dr. CL.
- PN is sleeping alone, as is DD.
- GY is still washing her feet regularly, and is still sleeping with FP.

---

embryo development, and fractal mathematics, is turning his attention to yet another aspect of growth: miniature roasting chickens. Dr. Natalee, with funding from a large commercial producer of poultry and poultry products, is working on new ways to grow teeny-tiny tasty Cornish hens.

Dr. Natalee's career has been marred by unfortunate calamaties. In 1978 he was nearing completion of a revolutionary new optical telescope on a mountaintop in Talca, Chile. Dr. Natale at the time was studying the growth of galaxies. The new apparatus was based on incremental mirror technology, and showed promise of increased magnifying power and lower construction cost than is possible with traditional designs. The project was ended by a sudden accident when a fragment of Skylab fell on the telescope and demolished it.

CHAPTER 4

TEACHERS VS EDUCATION

POSITIVE REINFORCEMENT OF POSITIVE REINFORCEMENT: JUST SAY YES

Linder Krim-Traven, Ed.M.
Chicago, Illinois

Abstract

A survey of educational authorities confirms that positive reinforcement is the only valid teaching method. The survey also confirms that there are no pedagogic underpinnings for the layperson's belief that educators should correct students' mistakes.[1]

Rationale

This study was undertaken to settle a long-standing dispute between professional educators and laypeople.

Educational authorities have long contended that the only valid way to teach anything is to use positive reinforcement. Positive reinforcement is the practice of rewarding a student immediately each time the student correctly answers a question or passes a test. The reward can be money or a small gift; alternatively, the reward can consist of effusive praise from the teacher.

Laypeople, on the other hand, typically believe that "people learn from their mistakes" and that when a student gives a wrong answer, the student should be told that the answer was wrong and should even be encouraged to discuss why it was wrong. This practice, which has long been known to damage students' self-esteem, is what educational authorities refer to as "negative reinforcement."

Methodology

We surveyed 400 accredited educational authorities[2] and 400 laypeople. Using mathematics to assess the numeric basis of our work, we applied advanced arithmetic and statistical techniques. We made intensive use of the mathematic techniques of subtraction, addition, and percentages.[3]

The survey consisted of two questions:

Question #1: Should teaching be based entirely on positive reinforcement?

Question #2: Should students be informed of their mistakes?

Results

Question #1
Educational authorities:
   82 percent[4] responded affirmatively.[5]
   12 percent[6] responded negatively.[7]
Laypeople:
   3 percent answered affirmatively.
   97 percent responded negatively.

Question #2
Educational authorities:
   12 percent responded affirmatively.
   88 percent responded negatively.
Laypeople:
   96 percent responded affirmatively.
   4 percent responded negatively.

LETTERS TO THE EDITOR

A Dissenting View

To the Editor:
No. No. No. No. No. No. No. No. No. No. No. No.

Jariel Ed Minitor
Grover Economic Research Foundation
Washington, D.C.

Interpretation

Our survey of educational authorities demonstrates that teaching should be based entirely on positive reinforcement.

Comments

Test subjects—whether humans, rats, or educational authorities—yield remarkably consistent results. Positive reinforcement, whether in the form of gifts, sugar water, or other accepted methods, is effective. Our negative reinforcement survey confirms that subjects do not "learn from their mistakes."

Implications of the Study

The author is highly disturbed, especially by those portions of the study that concern laypeople. It is difficult to escape the conclusion that laypeople are anti-intellectual and that they constitute an extremely serious threat to the quality of teaching in our society.

Laypeople present a dangerous challenge to the pedagogic underpinnings on which educational authorities rely. Laypeople completely misunderstand education. Ironically, *they* do not learn from their mistakes.

---

1. In other words, it is pedagogically and psychologically harmful to make students aware of their mistakes.

2. Holders of the Ed.M. degree and/or the Ed.D. degree.

3. Foster RF. *Arithmetic for Graduate School.* Wonder Press, 1989, pp. 1–18. Foster makes the point that negative numbers must always be preceded by a negative sign. Moreover, he demonstrates convincingly that nonnegative ("positive") numbers can, under some circumstances, be written without a plus sign.

4. Percent can also be written with the % symbol. The author chose to use the long form to ensure maximal mathematic rigor.

5. These respondents answered the question by saying "Yes."

6. This number was derived by subtracting two other numbers. (See Foster,[3] pp. 56–64, for more detail about the mathematics involved.)

7. These respondents answered the question by saying "No."

TECHNOLOGY UPDATE

An inside glimpse at what's new in emerging technologies

Stephen Drew

The Drain Brain

The world has been waiting for an intelligent garbage disposal unit. The promised wasteland may now be within sight.

Winchester Digital Systems has demonstrated a prototype unit that it calls the "drain brain." The drain brain detects the presence of indigestible items such as pots, silver spoons, rutabagas, and eyeglasses. It identifies and rejects items on the basis of their acoustic signatures, rather than their chemical composition, electromagnetic properties, or tactile or visual analysis. Wags have dubbed the technology "garbage wars."

The unit itself is small but includes a surprising amount of miniaturized (primarily VLSI) acoustic and electrical equipment. "We supply everything but the kitchen sink," joked Michael Sussman, the firm's founder and principal scientist. Sussman says the drain brain technology is more reliable and far less expensive than competing approaches that use "Star Wars" technology to identify and destroy incoming objects.

The drain brain does not attempt to destroy indigestible objects. Instead, it ejects them. Sussman points out that "this can be a substantial advantage if you've accidentally dropped valuable jewelry, coins, or plutonium into the sink."

The technology is not quite ready for market, however. Sussman says that "one problem is it doesn't work well for fingers. Early tests show that by the time you detect the acoustic signature of the finger, it's too late."

Winchester Digital's new quality control engineer, Adam Baum, is confident that the problem can be solved. "It's only fingers that seem to cause trouble," he said. "So if we have to, we'll just switch to analog technology."

PECULIAR RELATIONSHIPS BETWEEN AUTHORS AND THE SUBJECTS OF THEIR STUDIES

Alexander Kohn, Ph.D.
Anthropobibliographic Institute,
Tel Aviv

I. Preliminary Study

It has often amazed us how a scientist's choice of subject is seemingly associated with his or her name. A short review of the pertinent literature will demonstrate this point.

Lord Brain recently reviewed the brain mechanism and models[1]; together with Head he gave his ideas about *The Man and His Ideas.*[2] Some Foxes dealt with rats and dogs: One Fox studies the effect of trypan blue on rat embryos,[3] and another Fox psychoanalyzed dogs.[4] Harm showed that trypan blue was harmful to rabbit embryos.[5] Born *et al* measured the changes in the heart and lungs at birth.[6] It is a little strange that Bacon should have studied sugars in the blood of sheep.[7] Quite understandably, Amoroso was interested in the endocrinology of pregnancy,[8] and Seegal in immunofertility.[9]

For people interested in tissue cultures and religion we recommend the paper of Pious and Hamburger, who studied 50 cultures of human foreskin cells.[10] Data from Price were used to compute the values.[11] For botanists, we suggest Pond's paper on aquatic plants,[12] and for those interested in female anatomy, the paper of Goodheart on toplessness.[13]

Our attention was attracted also to some books: Dull and Dull wrote a book of mathematics for engineers,[14] Glasscock is the author of isotopic gas analysis for biochemists.[15] Biology of the laboratory mouse was written by C.C. Little,[16] while Smaller dealt with the structure of biomolecules (which are smaller still).[17] It is surely a coincidence that Hand had to do with man-

agement of bilateral undescended testes,[18] and that Professor Fleisch wrote a book on proper nutrition.[19]

II. Choice of Collaborators

We find that the belletristical value of a scientific publication is much enhanced by a proper choice of a collaborating author. Some examples illustrate this point.

The case of Brain and Head dealing with ideas of man was already mentioned.[2] Other combinations are Ham and Plate;[20] I.M. Tough, Brown Court, and King[21]; German, Bird[22]; Grey, Mutton *et al*[23]; Holland, Doll[24]; and Chu and You.[25]

Authors who prefer to stay single may have good reasons for doing so. For instance, it would be embarrassing for Poor[26] and Fortune[27] to appear together, or for the collaboration of Sell[28] and Favour[29] to make a favorable impression.

1. Lord Brain. Brain mechanisms and models. *Nature* 1964; 203:3.

2. Brain R, Head H. The man and his ideas. *Brain* 1961; 84:561.

3. Fox HM *et al.* Effect of trypan blue on rat embryo. *Proc Amer Assoc Anat* Buffalo, New York, April 2, 1958.

4. Fox HW. A sociosexual behaviour abnormality in the dog resembling Oedipus complex in man. *J Am Vet Med Assoc* 1964; 144:868.

5. Harm H. Der Einfluss von Trypanblau auf die Nachkommenschaft traechtiger Kaninchen. *Z Naturforsch* 1964; 9b:536.

6. Born GVR *et al.* Changes in the heart and lungs at birth. Cold Spring Harbor Symp 1954; 19:102.

7. Bacon JSD. Fructose and glucose in blood of fetal sheep. *Biochem J* 1948; 42:397.

8. Amoroso EC. Third International Rheumatology Congress, *Brit Med J* 1955; ii:117.

9. Seegal BC. *Symposium in immunofertility.* La Jolla Population Council, 1962, p. 215.

10. Pious DA, Hamburger RN, Miles SE. Clonal growth of primary human cell cultures. *Exp Cell Res* 1964; 33:495.

11. Price WC. Thermal inactivation rates of four plant viruses. *Arch Ges Virusforsch* 1940; 1:373.

12. Pond RH. *The biological relation of aquatic plants to the substratum.* Report of the US Fish Commission, 1901, pp. 483–525.

13. Goodheart CB. A biological view of toplessness. *New Scientist* Sept. 3, 1964, p. 558.

14. Dull RN, Dull R. *Mathematics for Engineers*. 3rd ed. New York; McGraw Hill, 1951.

15. Glasscock R. *Isotopic Gas Analysis for Biochemists*. New York; Academic Press, 1954.

16. Little CC, *Biology of the Laboratory Mouse*. New York; Dover, 1956.

17. Smaller B. In: Duchesne ed. *Structure of Biomolecules*. New York; Wiley, 1963.

18. Hand JR. Management of bilateral undescended testes. *Postgrad Med* 1963; 33:480.

19. Fleisch A. Erhaehren wir uns richtig. Thieme. Stuttgart; Springer-Verlag, 1961.

20. Ham JS, Plate JR. *J Chem Phys* 1952; 20:335.

21. Tough IM, Brown Court, King MJ. *Lancet* 1962; ii:335.

22. German L, Bird DE. *Lancet* 1961; ii:48.

23. Grey JE, Mutton DE, Ashby AV. *Lancet* 1962; i:21.

24. Holland WW, Doll R. *Brit J Cancer* 1962; 16:177.

25. Chu JP, You SS. *J Endocrinol* 1946; 4:392.

26. Poor E. *Transplant Bull* 1957; 4:143.

27. Fortune DW. *Lancet,* 1962; i:537.

28. Sell KW *et al.* Research in burns. Am Inst Biol Sci 1962, 351.

29. Favour CB. *Ann NY Acad Sci* 1958; 73:590.

NOBEL THOUGHTS

Profound insights of the laureates

Marc Abrahams

Henry Kendall is the Julius A. Stratton Professor of Physics at MIT, and is a co-founder of the Union of Concerned Scientists. In 1990 he was awarded the Nobel Prize in Physics for his work in obtaining experimental evidence for the existence of quarks. He spoke, via telephone, from his office in Cambridge. Some of Professor Kendall's doodles are shown at right.

How much time do you spend daydreaming?
A fair amount.

When you doodle on paper, what does it look like?
The doodling is usually done under circumstances where it distracts me from something that's extraordinarily tedious or tiresome. I try frequently to do projections of three-dimensional structures because it can take my mind off of inconsequential events better than doing plain two-dimensional structures.

Do you recommend that people whistle while they work?
It depends on the circumstances. I regard it as infinitely superior to smoking. It's less stress-relieving than having a drink during work. And it's much less stress-relieving than having several drinks. But the latter therapy has to be considered on a case-by-case basis with due regard to the circumstances.

Do you have any advice for young people who are entering the field?
Well, I would say that my previous answers reflect many years of participation and interest in the field. But my experience does not constitute an adequate base for advising young people at the present time.

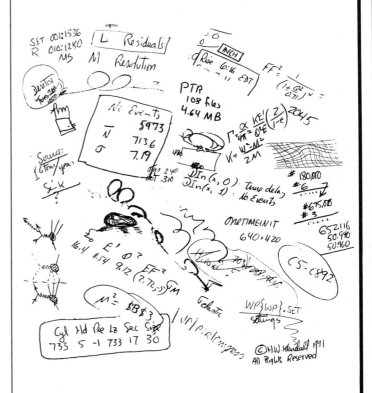

THE RELATIONSHIP BETWEEN SURNAME INITIALS AND ACADEMIC SUCCESS *OR* WHY WAS AABEL APPOINTED PROFESSOR?

Stephan Rössner
Stockholm, Sweden

The recent study by Eagel concerning the relationship between surname and cancer[1] underlines the importance of a research field that has received little attention so far, namely, the importance of the surname initial. Eagel demonstrated convincingly that some individuals with certain surname initials run a considerably greater risk than others of developing cancer. Among the high-risk group were those unfortunates with surnames beginning with M, K, S, D, and Z, whereas the initials J, A, and E offered protection. The present author (Rössner) was relieved to see that R offered a slight, albeit insignificant, degree of protection.

The question of alphabetic order, or rather disorder, is of utmost importance not only in identifying clinically important predictors for severe disease but also in elucidating the mechanisms that explain why some of us succeed in academic careers while others fail.

I was struck by the fact that in Sweden, 64% of all professors in internal medicine had surnames beginning with the initial B. This observation prompted the following questions: Is it possible that these scientists are promoted because they are repeatedly found as (alphabetically) first authors on papers emerging from their departments? Will their role as leaders in science be overemphasized because the principles, syndromes, equations, tests, methods, or whatever they have de-

veloped will carry their name, whereas the poor collaborating Smiths and Watsons (not to mention Rössners) will be forgotten?

The present study was carried out to analyze this question. The surname initials of all professors and associate professors of the Karolinska Institute were compiled from the faculty directory. On the assumption that these distinguished scholars, surprising as it may seem, were once medical students, the catalog of the Medical Students Association was analyzed in an identical fashion. The distribution of the surname initials of these groups is shown in Table 1, which demonstrates that

Table 1

Swedish Professors and Their Surname Initial

Initials	Professors	Associate Professors	Medical Students
ABC-LMN	76.3%	71.6%	66.8%
OPQ-ZÅÄÖ°	23.7%	28.4%	33.2%

°Ö is the last letter in the Swedish alphabet.
$p < 0.05$.

professors (who probably should be considered the most successful) have a significantly higher proportion of surname initials from the proximal part of the alphabet than associate professors (slightly less successful), who, however, are still over-represented in the A-N region in comparison to the students.

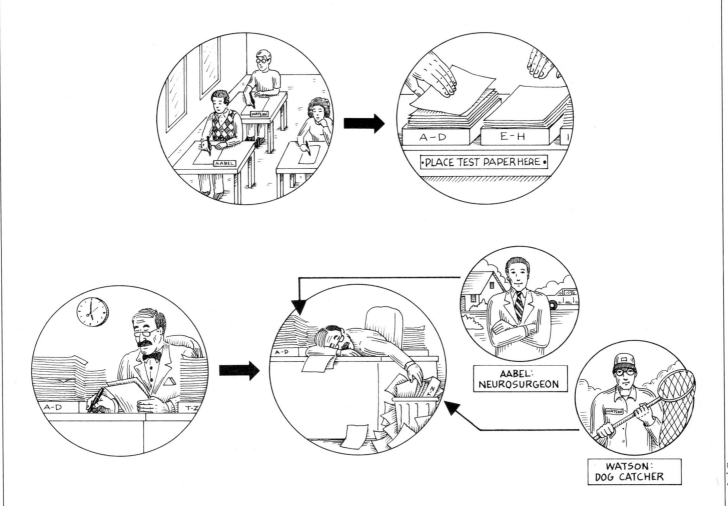

Jack Tom

This hardly comes as a surprise to many of us. In fact, in performing a search on the importance of alphabetical order on research careers, several authors were found to have addressed the question. It has been shown that authors with surnames from the distal part of the alphabet underpublish to a significant degree in journals that insist on listing authors in alphabetical order. The frustration of always being last in school, military service, medical examination, and vaccination caused by the fact that you are a "Z" person may damage one's ego for a lifetime. A study by Weston claimed that people with surnames in the end of the alphabet have significantly shorter life-spans than others. A subsequent analysis of the obituaries of more than 500 individuals published in the Salt Lake *Tribune*, however, could not confirm this hypothesis.

In medical school, individuals from the first part of the alphabet do better. Several interesting findings emerge from a recent study by Pritchard[2] concerning results in undergraduate studies of genetics. First,

women do significantly better than men, which might not come as a surprise to some of us. Second, the failure rate was considerably higher in students way back in the alphabet. Males in the A-C region had a failure rate of 14%, whereas those in the T-Z region had a failure rate of 33%. Distinction awards were given to 50% of A women and to only 22% of T-Z women. Pritchard considers the differences too high to be ignored and concludes that students at the front of the alphabet are different. This position affects the concept of self-worth and hence the later level of academic performance.

All this is important to know when you have had a bad day, scientific manuscripts rejected, promotion postponed, and realize that you are an R, like Stephan Rössner.

1. Eagel BA. What's in a name? The relationship between surname and cancer occurrence. *Jour Irr Res* 1988; 33:9–10.

2. Pritchard DD. Effects of sex and alphabetical listing on examination performance of medical students. *Med Ed* 1988; 22:205–210.

IN MEMORIAM: DR. ELROD HIBBIRD
Founder of Dianautics—The Science of Flying Brains

Stephen Drew

The man whose theory of flying brains revolutionized science, religion, and international banking practices is dead. Dr. Elrod Hibbird succumbed to spontaneous combustion, an ailment from which he had suffered chronically, in an alleyway behind his Church headquarters in Oslo, Norway. He was 83 years old.

When Elrod Hibbird was a young man working in a factory in Cincinnati, Ohio, he was assigned to write the instructions for a new model of vacuum cleaner. The experience changed his life and inspired him to found a new religion. Hibbird felt that he had seen, in the machine's irresistible gathering of gray matter, a key to understanding how human intelligence developed from billions and billions of bits of grimy interstellar matter.

Human minds, Hibbird theorized, are actually distributed throughout the cosmos. Therefore, he concluded, human bodies are illusory, and actions performed toward another person are merely "dream-fictions," for which one need not be held morally or legally accountable. Hibbird quit his job at the vacuum cleaner factory and devoted his life to this idea.

Hibbird's 1948 book, *Dianautics—The Science of Flying Brains*, became a Bible for teaching the theory and practice of what Hibbird called "brains at a distance." Together with several followers, and with financial backing from the venture capital firm of Homburg and Inqvest, Hibbird went on to found the Church of Dianautic Science.

In ensuing years the Church grew to include, by its own count, more than 1.4 million members in 65 countries and 335,921 galaxies. From its earliest days the Church has drawn its membership heavily from the scientific, engineering, and show business communities. Its annual revenues are estimated at more than $3.2 billion, with real estate and other capital holdings thought to be worth approximately $24.6 billion.

Hibbird progressed through a series of experiments with his new science. He arranged for flying brains to run mazes at a distance. Flying brains were sent to do research inside particle accelerators. Later, Hibbird used flying brains to actually carry out gedanken experiments—thought experiments—including one based on Einstein's well-known twins paradox. Einstein had mentally calculated the relativistic effects, in time and space, on two twin brothers, one of whom took a round trip on a speedy spaceship while the other remained on earth. Einstein believed that one of the brothers would grow older than his twin. Hibbird's flying brain experiment showed this not to be the case.

Hibbird was an intensely private man. Throughout his life he declined to reveal details about his family and upbringing. According to Church of Dianautic Science officials, Hibbird's passing is "illusory, as are all things human." Previous publications from the Church indicate that Hibbird is survived by several wives, children, grandchildren, and a variety of "granular beings."

A 1971 photo of the late Elrod Hibbird. Photo: B. Vanatian

An inside glimpse at what's new in emerging technologies

Stephen Drew

Intelligent Liquids

During the past three years, much progress has been made in the development of intelligent liquids. "Smart water," as the new substances are coming to be known, is being adapted for use in several fields. The research, conducted at Kurwell Aqua Systems, Inc. in Phoenix, Arizona, is largely classified, but it is widely believed that next year will see the first commercial applications reach the marketplace.

As an ingredient of soft drinks and other foodstuffs, smart water is able to sense the chemical environment of the mouth and stomach, and adjust its pH value accordingly. Kurwell chemists have demonstrated a smart-water–based flavoring additive whose taste resembles that of anise, a flavor that some people find delightful and others mildly unpleasant. The Kurwell smart anise is able, in at least some cases, to understand when the host person dislikes its initial taste, and to rapidly produce a sweet, venison-like taste that most anisephobes find more palatable.

In hydraulic microcontrol mechanisms, smart water is proving to be invaluable for its ability to alter its viscosity in response to changing conditions. Tests have shown some indication that it may be possible to develop forms of the liquid which are able to substantially increase their viscosity when a leak develops in the pipe that contains them. Kurwell researchers are, for example, working on a smart water antifreeze fluid for automobiles; the goal is to have a fluid that protects an engine against extremes of both heat and cold, and that is also able to plug radiator ruptures. (Kurwell has already licensed the rights to one spinoff of this technology—a variable-viscosity "smart ketchup.")

The wine industry is known to be looking into the technology. Kurwell's marketing officials only half-jokingly allude to the prospect of creating a "truly sophisticated" wine. Kurwell is also known to be negotiating development and licensing arrangements to adapt smart water technology to a variety of more broadly useful agricultural applications, including an irrigant additive that, researchers speculate, could alleviate the onset of frost in citrus fruits and other crops that are especially temperature-sensitive.

Kurwell is also doing experimental work in pharmaceuticals. The company has achieved preliminary results involving a smart water laxative. Kurwell says it is unable to release its results at present, but predicts an imminent breakthrough.

Kurwell's smart water may help answer the long-standing question: Is the glass half empty or half full?

THE EVOLUTION OF COLLEGE TEXTBOOKS

S.A. Rudin
Charleston, West Virginia

Since I like to read the classics, I went into the bookstore of a small college in West Virginia—the only political entity governed by principles adumbrated by Sigmund Freud, after he stopped his research on cocaine, but before he invented psychoanalysis—and asked the salesclerk if they had any Penguin books. She said that they did not, but that they had some on pelicans and seagulls. Before wending my way out, I glanced over the assembled tools of enlightenment.

There was a psychology text, fully 14″ wide, 10″ high, and 4.5″ thick. The margins were a full 5″, allowing only a pitiful column of text 4″ wide to dribble down the center of the page. Every page had something other than text on it; a photo, a drawing of a piece of apparatus with a wretched rat gazing dumbly out of it, a cartoon from the *New Yorker,* or a photo of one of the author's children cutely throwing a Stanford-Binet at the camera. A political science text had photos of the U.S. Congress, a recent president of the United States giving a press conference, and other scenes of death and destruction. The sociology book had pages of photos of cops beating up a one-armed woman named Lin Pao Sanchez; the chemistry book had perverted drawings of two-valenced elements trying to combine with three-valenced elements. The calculus book showed what mathematical symbols did by drawing them on road signs. And one on computer programming illustrated a loop with a dozen dogs interspersed among a dozen cats, all chasing each other's tails.

How did we come to this pass? An examination of the development of college textbooks is in order.

The first textbook was written by Claude, le Comte de Cromagnon, at Chez Cavitie, France, on the walls of the family cave. It dealt with how to sneak up on deer, the proper means of preparing cave bear soup, how to bury your flake culture arrowheads 40 meters underground to drive future archeologists crazy, and the proper incantations to Zorkon and the National Rifle Association, gods of the hunt. Other incantations included those to the goddess and god of fertility, Lysistrata and Testicles.

Isidore Aleph, Seymour Beth, and Irving Gimmel later invented the alphabet. Unemployed scribes, they were lying around idling; their Babylonian beer had gone flat (*"Drink old Babylonian/Feel like a ziggurat"*); and they were angry, since King Cuneiform XVIII had put a tax on clay tablets. This invention led to further writing on papyrus, until Marco Polo came back from China with paper. After that, publishing boomed.

Things scrolled along pretty well, but not in quantity. Then one day, Hugo Gutenberg, dwarf brother of Johann, invented the paperback book so he could have something that fit easily into his hands. At last, everyone could be educated. Two American inventions funded this development: the G.I. Bill and the Cold War, which also paid a lot of graduate students' way through graduate school.

Unfortunately, art directors, media consultants, and market researchers decided that textbooks needed to be enlivened. For years, books had been published with plates—engravings—of the insides of dead creatures with nauseating diseases, fields of stars (usually indistinguishable from scratches on the plate), and profiles of German scientists with bristling Prussian mustaches. What with television as the main medium of discourse for most of the population, pictures and cartoons were multiplied a couple of googolplexes.

On a recent visit, Harold Buchtten, publicity director of Communitex, the world's largest publisher of college textbooks, displayed the very latest in college textbooks. Cartoons abounded, photos were shown in a centerfold or two, and drawings covered everything. Only three things were missing: words, letters, and numbers. Here, clearly, is the ultimate college textbook.

IN MEMORIAM: RAUL DE WOMYNN
The Father of Desensification

Stephen Drew

To some, Raul de Womynn was a consummate scoundrel. To others, he was a liar, a cheat, a bigamist, a Fascist quisling, a monstrous villain, a vile rogue, a snake-oil salesman, a Jerry Sneak, a charlatan, a mountebank, a saltimbanco, a medicaster. The beloved chairman of New Haven University's Literature Department was the inspiration for, and leader of, a generation of literary theorists.

Raul de Womynn and his followers reinvented the study of literature. Their banner was an intellectual theory known as desensification. Their slogan was "Stop Making Sense." They showed scholars that the most powerful way to criticize a piece of text is to read the words in random order. For thousands of scholars who despised the love of books, de Womynn's death leaves a gaping ache.

The details of de Womynn's early life provided ammunition for his enemies in academe. Five years ago, a rural French newspaper revealed that, as a young man, de Womynn served as a speechwriter for Benito Mussolini, Joseph Stalin, and Adolph Hitler; that he sold obscene postcards to children; that he embezzled money from an orphans' fund; that he smuggled drugs into France, Belgium, and Italy; and that he committed a string of grisly murders. He also married and abandoned 17 women in 13 countries before emigrating to the United States in 1954, where he married and abandoned another 17 women.

Raul de Womynn's colleagues and admirers were heartened by the charges against their master. Using the desensification techniques that de Womynn had taught them, they explained that history is a random confluence of irrelevant events that carry no significance whatsoever. Professor de Womynn, they concluded, was really the twentieth century's archetypal victim.

As de Womynn himself often said, "History is ultimately a fiction, a predicament of language, a *joie d'esprit* that is subject neither to sensible interpretation nor to moral judgment." Thus, his supporters say, by committing unspeakable atrocities, de Womynn became the embodiment of random goodness. A generation of literature students agrees.

Raul de Womynn died, at the age of 74, in a style consistent with the tenets of desensification. He was gunned down by New Haven police while committing armed robbery at a convenience store. The Modern Language Association plans to hold an annual commemorative seminar at the site.

Raul de Womynn, New Haven, 1987.

HOW TO MAKE A SCIENTIFIC LECTURE UNBEARABLE

X. Perry Mental

Ness Ziona, Israel

At a symposium, meeting, or congress when there are a number of speakers, there comes a moment when your name is called. A nice ploy to attract the audience's attention at this stage is to place yourself in the middle of the last row, so that when you are introduced, you raise the whole row, step on their toes, proceed slowly to the front of the hall, and then start searching your pockets for a convoluted pack of lecture notes. Next you extract from another pocket a package of slides with which you go back to the projectionist and enter into an animated discussion with him trying to explain which slide is first and which side up and instructing him, "And don't forget to show slide 3 again after slide 7." Then you go back up to the lectern, and start searching for your reading glasses. (They probably are in an unex-

pected pocket.) Next you proceed to "read the paper," and we mean literally "read" it. This technique of delivering a lecture is defined by Prof. Sabin as "kissing over a telephone—completely tasteless."

If you wish to put your audience to sleep as soon as possible after starting to lecture, begin with the enumeration of all historically important papers published in the last 50 years that have any bearing on the subject matter. Another well-tested method is to start talking about something that has nothing to do with the subject by saying for instance: "Before we turn to the discussion of . . . , let us shortly review . . . , etc."

Beginning at the beginning is an unpardonable mistake. Some speakers use the so called multiple colon technique. They say: "Mme. Chairman, I should like to say: The situation is as follows: I mean to say that: I should like to clarify in this lecture some points which are not sufficiently clear: etc., etc." If you continue for a few minutes in this vein, you quickly lose the audience.

Jack Tom

A useful habit to distract the attention of the audience is to have a "tic," like twitching one cheek, a sniffing movement of the nose, twisting of the neck, and buttoning and unbuttoning your jacket, etc. Putting on and removing the reading glasses while you talk, and glancing at the audience, can sometimes replace such a tic. If you manage to combine the tic with the glasses, so much the better.

Some sophisticated speakers like to introduce quotations in their lectures. Shakespeare, Einstein, and Wendell Holmes are quite safe in this respect. The trouble begins when the quotations are from classical literature or the Bible in the original language, be it Greek, Latin, Hebrew, or Sanskrit.

We shall not go into the use of slides in the lecture. This subject has already been amply discussed by Wilkinson (*JIR* Vol. 10 #1). Suffice it to say that keeping the audience in the dark while running through some 50 or 60 slides (preferably representing complicated tables or figures), will greatly assist the listeners in taking a solid nap. We shall not discuss the obvious ploy of having the slides intentionally inserted the wrong way, so that it takes the projectionist six or eight trials before they are correctly set. If you need some breathing space as a speaker, this is how you get it.

Now as to elocutionary technique. Wrongly adjusted microphones help in losing the audience. This is especially true if there is only one microphone on the speaker's table and if he happens to wander around while pointing to the screen or writing on the blackboard. If you happen to be attached to a neck or breast microphone via an umbilical cord, then a good method is to stand in front of the blackboard with your back to the audience, and speak over your shoulder so that the microphone is well-screened by your shoulders. Lip-readers in the back row might understand what you are saying.

Some speakers like to doodle on the blackboard while talking. We have in our files a collection of such doodles and a dictionary, which helps to understand what said doodles mean.

Now most of the slides are over, most of the audience is nicely asleep, and you are near the end of your lecture. An hour has gone by, and you find that you barely managed to convey half of the material you intended; you notice that your chairman is fidgeting and tries to catch your eye to indicate that you must end. You then say: "In conclusion, I should like to say . . . ," which gives you some 5 or 10 minutes' grace. If you cannot end by then say, "Finally, these results indicate, etc . . . ," which give you another few minutes, when you still may say, "To sum up, . . . etc."

There are speakers who, when warned by blinking lights or threatening posture of the chairman, put themselves at the mercy of the chairman and ask, "How much time do I have?" which does not give them really a chance if the chairman is not a mouse. Some bolder types usually say, "If I have still 5 (or 10) minutes, then . . ." and ignore the chairman altogether. Some intimidated speakers accelerate the rate of their delivery to a speed that permits only highly trained experts to keep track of the subject matter.

It has been suggested that the listeners should organize themselves in a Society for the Prevention of Cruelty to Listeners and present the speakers with rules, regulations, and sanctions before they start talking.

SPECIAL BOOK EXCERPT

(First of a two-part series)

How Executives Overcome the Fear of Reading Books
Ipthog T.C. dePrieze, M.D., J.D., M.Ed., D.D.
Institute Professor of Commerce and Learning, The University College Institute
New York, New York

Section 1: Why should I read this book?

This book will help you overcome the fear of reading books. Many executives suffer from this fear.

Section 2: How should I use this book?

This book is organized in tiny, comfortable sections. Read this book beginning with the first section. Proceed to the next section. Repeat this process. When you reach the end of the book, stop reading.

Section 3: I do things my own way. However, I am curious to know how other executives overcome their fear of reading books. How do other executives overcome this fear?

Executives overcome their fear of reading books by reading this book. They begin by reading the first section. Then they proceed to the next section. They repeat this process. When they reach the end of the book, they stop reading.

Styles, trends, and tidbits culled from leading research journals

by Alice Shirell Kaswell

O ver the past few years, the technology for manipulating matter and time on a vanishingly small scale has advanced at an astonishing pace. The research journal the **New York Times** has regularly been reporting on the startling development, by Origins Natural Resources, Inc., of zero oil. The oil contains redness-fighting Cola-Nut. The **Times** (November 25, 1990, sec. 6, p. 25) also reported on Origins' introduction of an original formula for pressed powder that is softened and "sheered" with nature's minerals.

Time Zone Eyes

Much progress is being made on the manipulation and isolation of time. The 1990 issue of the journal **Ecco Bella** identifies 10 colors that are, in fact, timeless. (The same issue also contains an unrelated—but fascinating—report indicating a correlation between the purchase of wonderful snacks in beautifully packaged tins and the sustainable use of rainforests.) The October 18, 1990, issue of the **New York Times** contains a report (p. 15) about Elizabeth Arden's Ceramide Time Complex. The November 4 issue of the same journal reports (pp. 26–27) on Estée Lauder's Time Zone Eyes Ultra-Hydrating Complex, a liqui-creme formulation that drives ultra-hydrators into skin, and which is noncomedogenic. The November 1990 issue of **Scientific American** (vol. 263, no. 5) contains, coincidentally, a report (p. 127) on polymers that conduct in one dimension.

Arts and Sciences and Gel

"Society is just beginning to recognize the role grooming plays in self-esteem and individual accomplishment," Nexxus president Stephen Redding reports (p. 38) in the November 1990 issue of the research journal **American Salon**. Redding is donating $10,000 to help found the Museum of Cosmetology Arts and Sciences. (To make a tax deductible donation, contact Diane Sherrill White, the museum committee chair, at 704-873-3010.) Other valuable reports are presented in the same journal. One of them (p. 45) details how researchers have developed a transparent liquid gel that goes all the way to a new dimension in resurfacing. Another report (p. 51) reveals news that regular readers of this column

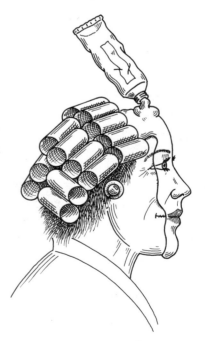

have been expecting: that Thymoderm's Thymus Extract has been hailed as a major breakthrough. Matrix has been awarded a copyright for the phrase "Science in sync with nature." On page 55, investigators describe that research institute's Systeme Biolage, a harmonious

Jack Tom

blend of pure biologic extracts and scientific ingredients . . . *American Salon* also contains a report (p. 72) about L'anza's new generation reconstructor that moisturizes while replacing protein.

Parameters, Traveling Objects, and More Gel

Section 6 of the November 11, 1990, issue of the *New York Times* presents an array of salient results. Investigator John Russell shows convincingly (pp. 56–62) that the remarkable thing about the English painter Hodgkin's achievement "is that its parameters were established almost before he was out of the nursery." Investigator Nancy Newlove describes (p. 11a) a new class of traveling objects, including a traveling shirt and a traveling tie. There is also an analytical report (p. 15) about Clinique's synthesis of a lightweight gel with new-tech speed and skill.

Vital Elements

Investigator Melissa Bedolis, in the research journal *Nexxt* (1990, vol. 2, no. 2), describes Nexxus's attempts to infuse vital elements for color infusion. A series of experiments with Nexxus Botanoil Certified Oils and Nexxus Ensure produced hair with dimensional color. Moreover, the dimensional color lasted longer and processed faster.

Goats, Heartburn, Mouth-Chewing

The research journal *Glamour* (December, 1990) presents its recommendations (p. 42) of the circumstances under which nipples are to be coated with petroleum jelly, then covered with a gauze pad before being covered with a brassiere. The same journal reveals (p. 36) that Christian Dior mascara parfâit is formulated from the coat of cashmere goats, and (p. 32) presents a convincing argument that Neutrogena moisturizers are, in addition to being noncomedogenic, so pure that they work with skin. Investigator Stephanie Young finds (pp. 48–63) that fatty or very spicy foods can trigger heartburn, that nicotine is a stimulant, that alcohol increases urine production, and that crackers, potato chips, and cereals stay in the mouth longer than jelly beans, caramels, and chocolate. Young also reports the case of a subject who chews the inside of her mouth until it is raw and bleeding.

Microblush, a Virgin, Television

L'Oreal's invention of microblush is reported (pp. 10–11) in the September, 1990, issue of the *Ladies' Home Journal* (vol. CVIII, no. 9). This substance makes it possible, according to the report, for a person to become totally transparent. The same journal also reveals (p. 13) that the Secret organization is in possession of a wide solid that has enough strength for a male; however, the

Catherine Lazure

solid is pH-balanced for a female. Investigator Corinne Clements reports (pp. 18–22) the case of a woman who is a virgin after 16 years of marriage, and who was able to confirm that fact by viewing television.

Elemental Vitality

Clinique has developed a system so noticeably effective that nothing else—before or since—has ever performed as well. This fact is reported in the November 5, 1990, issue of the research journal the **New Yorker** (p. 5). The journal also presents (p. 61) the case of a woman who takes nature into her bath.

Steamers, AI Kits, Cool Semen

Natural Life is the only formula with chelated minerals and vitamin C polyascorbate (ester-C), according to a report (p. 185) in the journal **Dog World** (December, 1990, vol. 75, no. 12). Mr. Christal's Australian Medicated Shampoo for Dogs and Cats contains salicylic acid and melaleuca alternifolia (Australian tea tree oil), and, according to a report on p. 136, is pH-balanced and contains no animal byproducts. Investigator Richard E. Bradley, Sr., M.S., D.V.M., Ph.D., recommends (p. 7) small propane-fired steamers designed for degreasing car engines. Investigator Dwight D. Bowman, M.S., Ph.D., concludes (also p. 7) that *Giardia* might be the new parasite of the decade, and believes it was previously ignored because it was considered a nonpathogenic commensal organism. A report on p. 1 of the journal finds that Eukanuba is optimum nutrition. Adult Eukanuba is a maintenance diet. Adult Eukanuba is the only food canines require. Adult Eukanuba contains ingredients. They are perfectly balanced. They are like chicken. They are like eggs. They are like rice. They are like corn. They are like top-grade poultry fat. Adult Eukanuba is efficient. According to a report on p. 178, AI kits and cool semen are now available from the Seager Canine Semen Bank. **Dog World** also presents a fresh report (p. 177) on Xoloitzcuintli.

Biomitism, Chassis, Nails

A startling report (p. 6) in the research journal **Cosmopolitan** (December, 1990) details how Lancôme Laboratories' Niosôme Daytime Skin Treatment is not a lotion or a cream, but a system of microscopic multilayered spheres that permeate the skin's surface layers with advanced microcarriers. The structure of the sphere itself is based on a phenomenon that Lancôme characterizes as "biomitism." Cover Girl has developed a liquid pencil (a report on p. 25 presents details) with a line that is pen-precise. The look is, according to the definition presented in the report, the line. Oil of Olay plans to change the way people wash their faces, according to research presented on p. 37. According to a report on p. 54, a Cray supercomputer-designed chassis makes a Sentra extremely rigid. Investigator Mallen de Santis collates (p. 58) empiric findings that ministers are low in testosterone. There is evidence (p. 126) that nails that are sculptured are more durable than nail tips. **Cosmopolitan** also reveals (p. 131) that doctors have recommended Advil™ over 50,000,000 times in 6 years, but does not make clear whether the years were consecutive, nor does it report as to whether the recommendations were followed. **Cosmopolitan** itself recommends (p. 98) that ladies-room visits be kept to under 5 minutes.

CHAPTER 5

THE UNBEARABLE COOLNESS OF FUSION

TABLETOP FUSION

Ponz Fleischedicher and Joseph Morton

Stanley University of Uganda, Pepper Lake City, Uganda

Abstract

The authors attempted to achieve tabletop fusion at room temperature (23° C). They succeeded.

Materials

(a) Two (2) tabletops, each composed of pasteboard and measuring 1 m × 2 m × 10 cm.
(b) A paste mixture composed of flour, water, and palladium grounds.
(c) Tritium, in trace amounts.

Method

The authors sprayed atomized paste between the two tabletops (Figure 1). The tabletops were brought into close contact. Pressure was then applied to their outer (noncontact) surfaces. The tritium was stored at a separate storage facility and was observed continuously during the course of the experiment.

Results

Tabletop fusion occurred (Figure 2). Trace amounts of tritium were observed at the storage facility.

Discussion

The authors have conducted this experiment repeatedly, each time obtaining the same results. Full details of the experiment will be published upon completion of routine administrative procedures and full registration of patent applications.

Acknowledgments

We wish to thank the many members of the press without whose guidance we would not have been able to achieve these results.

Figure 1

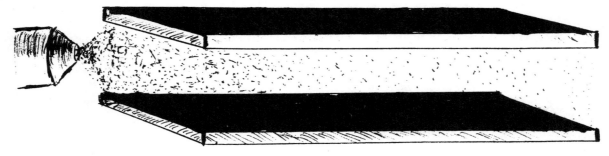

Figure 2

LUKEWARM FISSION

F. Stickney de Bouregas

New England Poultry Analytical Institute, Norwich, Vermont

Introduction

The recent observation by Ponzi and Fleischkopf of deuteron fusion in electrochemical cells suggested to us that discoveries of similar importance might be made by using metals other than palladium (Pd, Jun 166.20, +1.60). In particular we have focused our attention on tungsten (W).

Experimental Procedures

The experiments were carried out in an apparatus of the kind shown in Figure 1. The problem of confinement was addressed by using special cells or "jars" (designed by General Electric Co.). Instead of having an "anode" and a "cathode" separated by an aqueous (or polyaqueous) solution, each jar contained a single electrode, the *monode*, made of pure tungsten. A source of electricity was connected to the termini of the monode so that electrons could be galvanostatically compressed into the tungsten, creating a negatively charged plasma, according to the reaction steps:

$$W + e^- \ (115 \ eV) \quad We^- \qquad (i)$$
$$We^- + e^- \ (dto.) \quad We^{2-} \qquad (ii)$$
$$etc. \qquad\qquad (iii)$$

(Holes, which may be easier to compress than electrons, will be dealt with in a subsequent communication.)

The entire apparatus was placed in the basement, surrounded by yellow tape labeled "DANGER DANGER," to alleviate the anxiety of the students. (Students were in touch by walkie-talkie with the principal investigators, who were upstairs with the press.)

Results

EXPERIMENT 1

The galvanopiestic compression was continued for varying lengths of time. At the end of each run, the total photonic energy output was determined from the lexiphotometer readings; the time was determined by multiplicative photoergometry. The excess energy output appearing in the form of heat was calculated in some fashion by the computer. Representative results of these experiments are shown in Table 1. The following conclusions are immediately obvious:

Table 1. Generation of photonic energy E_{ph} (lexiphotometric) and excess thermal energy H_{xs} (Dosimetric) in W double-helix electrodes contained in jars of various sizes.

Size (cm)	Time (h)	E_{ph} (joule)	H_{xs} (joule)
8.0	1.0	1.4×10^5	1.0×10^5
	2.0	2.8×10^5	2.0×10^5
10.5	1.0	3.6×10^5	3.0×10^5
	2.0	7.2×10^5	6.1×10^5
	4.0	1.44×10^6	9.1×10^5
	8.0	2.82×10^6	2.0×10^6
	1023.5*	3.70×10^8	*

*We have to report that the last experiment spontaneously terminated with an ominous "pop."

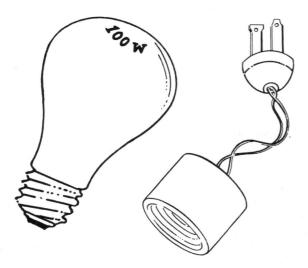

Figure 1. Experimental apparatus. The right-hand module was built in our workshops. Examples of the left-hand module, which could be replaced as required, were kindly supplied by General Electric Co. (Another version, having a different aspect ratio from the one shown, was provided by Phillips Gloueuijlaempijenfabruijken N. V.) Left: Source of electricity. Right: Lexiphotometer.

• Both the photonic energy and the excess thermal energy *increase with the size of the jar.*

• After the compression has been carried on for a sufficiently long time, *the total energy becomes very large and therefore must be attributed to nuclear reaction.*

EXPERIMENT 2

The meaning, if any, of the energy quantities depends on what is included and how various numbers are added or subtracted, a point that emerges clearly from the masterful theoretical discussion of Ponzi and Fleischkopf. A colleague proposed that we compare the energy outputs of Table 1 directly with *the total measured electrical energy outputs to the jar.*[1] Therefore we made multiple polyergometric determinations of input energy, as follows:

• The voltage E between the two terminals was determined potentiometrically by using a sample-and-hold circuit and A/D converter. The magnitude and sign of the voltage were found to vary considerably from one measurement to another, with all measurements falling in the range -168 to $+168$ volts. A statistical analysis was performed, and the mean voltage E was found to be zero, within 95% confidence limits.

• Similarly the current I was determined amperometrically. Within 95% confidence limits the current, too, was found to be zero.

• The power input P was then determined multiplicatometrically. It can be shown[2] that

$$P = E \times I = 0 \times 0 = 0 \quad \text{(iv)}.$$

The implications of this result, coupled with those given in Table 1, are astounding. Enormous amounts of energy—unexpectedly, more light than heat—are produced with *no measurable input of energy whatever.*

To determine more precisely the origin of these effects, we did further experiments that established the following facts: (1) Energy production does not occur until the electrode is connected to the source of electricity. It ceases when this contact is broken. (2) If the monode is removed but the rest of the apparatus is left intact, energy production ceases. (3) If the monode is replaced by one made of another material (spun nylon glass fibers, etc.) no measurable energy of any kind is produced.

JIR RECOMMENDS

*Articles, books, and other communications
that merit your attention*

Compiled by Stephen Drew, Norman D. Stevens,
and X. Perry Mental

"Elucidation of chemical compounds responsible for foot malodour," by F. Kanda, et al, *Br J Dermatol* 1990; 122:771. In one of the experiments described in this paper, 10 healthy adult male subjects exercised on an ergometer at 35° C and 50% humidity for 30 minutes. The five subjects who felt that they did have foot odor in fact did. The five who felt they did not in fact did not. (*JIR* thanks Alan Rockoff for this citation.)

Evidently, a source of electricity catalyzes some unknown kind of nuclear reaction, *which occurs only in tungsten monodes.* This reaction continues as long as the catalyst is present, ultimately yielding large and perhaps dangerous quantities of energy.

Nature of the Reaction

We performed analysis (using an instrument)[3] for elements in the mass range around 368, with negative results. Therefore we are not dealing with nuclear fusion. Therefore, the phenomenon in this case must be fission. To verify this assumption, we analyzed the gaseous material inside the jar after energy production had been occurring for some time; and found easily detectable amounts of the fission product ^{40}Ar.

Discussion

The observation of power generation from electrochemically compressed electrons in a W monode is a very surprising result, and it is evidently necessary to reconsider the quantum mechanics of electrons in such host lattices. We urge that no attempt be made to repeat these experiments because of their potential danger to our credibility.

Acknowledgments

We thank our coauthor for remaining anonymous.

---

1. This sounded risky, but we knew we didn't need to report any results we didn't like.

2. de Bouregas FS, *Anal Res*, to be submitted.

3. Instrumental analysis.

NEWSMAKER INTERVIEW: MARTIN FLEISCHMANN
Cold Fusion Pioneer

Marc Abrahams

In 1989, Martin Fleischmann and B. Stanley Pons startled the world. They announced the discovery of what appeared to be a new, easily attained form of nuclear fusion, called "cold fusion." Their work, carried out at the University of Utah in Salt Lake City, touched off a barrage of controversy.

Cold fusion, if it turns out to be feasible, would offer humankind the possibility of generating inexpensive, abundant, clean energy.

Critics have accused Fleischmann and Pons of a multitude of sins—of making claims unsubstantiated by their data, of publishing flawed data, or publishing insufficient information to allow other scientists to try to reproduce their experiments, and of publicizing their work via press conferences rather than by publishing it in peer-reviewed journals. At the same time, scientists at several institutions have reported some success in reproducing certain aspects of the research.

During the past two years, sensational rumors have swirled about Fleischmann and Pons. Now, in an exclusive interview with the *Journal of Irreproducible Results,* Martin Fleischmann reveals the truth about the recent rumors.

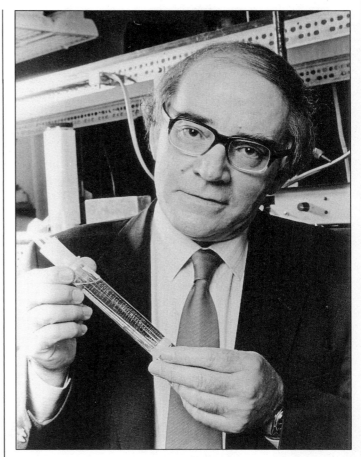

Q. Are the rumors true about you and Madonna?
A. No.

*Contains 100% gossip
from concentrate*

Compiled by Stephen Drew

Otolaryngology Recapitulates Phylogeny

Through the magic of home video, the general public is seeing the creation of a new scientist celebrity. Dr. Rachel Linley Wetmore of the Andover Clinic, who last year was the star of the unexpectedly best-selling videos "Aerobic Ontology" and "Physicians Recapitulate Phylogeny," has a new video that will be available in stores within the month. Called "Otolaryngology Recapitulates Phylogeny," it is already topping the sales charts in advance sales to video dealers.

Common Cold Fusion

In the United States, a teenager acquires a new sexually transmitted disease every 13 seconds (see the journal *Science News*, 9/15/90 for details; *Science News* does

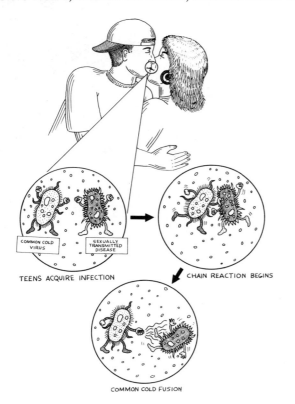

COMMON COLD VIRUS   SEXUALLY TRANSMITTED DISEASE

TEENS ACQUIRE INFECTION   CHAIN REACTION BEGINS

COMMON COLD FUSION

not specify the name of the teenager). Investigators have long been puzzled at how teenagers are able to survive this relentless bombardment of potentially fatal invaders. Now researchers have discovered a protective mechanism that may explain the mystery.

Michelle Greer-Scott and several collaborators at the Walden Genetics Institute in Santa Clara, California, have identified and documented a thiotimoline chain-reaction process that occurs spontaneously when certain viruses are present in high concentrations, and in sufficient diversity, in white cells. Greer-Scott reports that many of the viruses are also associated with influenza and the common cold.

The chain-reaction process is known as common cold fusion. It results in an excess of energy, in the form of heat. Following the reaction, the viral material is present only in the form of isolated nucleic acids and water. Greer-Scott says her group has also seen sporadic evidence of tritium production, but cautions that since tritium measurements are difficult to perform, the findings on this score are preliminary.

"Wonder Crack"

An intensely addictive new form of crack cocaine has been introduced in the U.S. and Europe in recent months, with an unusually insidious marketing campaign. Known as "wonder crack," the drug is reputed to have an unusual biologic property. Many of wonder crack's users believe that it helps prevent the buildup of arterial cholesterol, a major cause of heart disease. Drug dealers, and some holistic chiropractors, are claiming that the substance "builds strong bodies 12 ways." The drug may have some health benefits but it is more addictive than most other illegal drugs. Wonder crack is a highly refined form of crack cocaine that has been denatured with iwanuline-450, a chemical derivative of oat bran.

The View from Above

Spy satellites, prized for their role in surveying the earth's surface, may be replaced by a cheaper, simpler technology. It will soon be possible for an earth-based watcher to visually observe almost any other surface location instantly and at negligible expense.

Meteor burst communication, in which ground-based antennae bounce signals off the long, ionized trails left by meteors entering the upper atmosphere, is proving more adaptable than was first expected. Transmission of radio frequencies using meteor trails has long been understood. But last month, for the first time, researchers at Ogletech, Inc., demonstrated a prototype system that works in both the visual and infrared frequencies.

The demonstration took place at Ogletech's research facility in Teaneck, New Jersey. On a large-screen television monitor, reporters could clearly see, from above, bathers on a resort beach at Bimini in the Bahama Islands. The real-time images were extremely sharp, and showed an astonishingly high level of detail. Ogletech personnel treated the reporters to a virtual pore-by-pore examination of one beach-going couple.

SLEEP RESEARCH UPDATE

- SD is sleeping with NN, PDG, RL, FFD, and GW.
- DF is sleeping with LF, only to a statistically insignificant degree.
- LP now sleeps with DB.
- DB now sleeps with OG.
- OG now sleeps with WHF.
- QV now sleeps in the laboratory during weeknights and in the lunchroom on weekends.
- PT is sleeping with GF and LE.
- SG has received his doctorate, and is no longer sleeping with members of the Psychology Department.

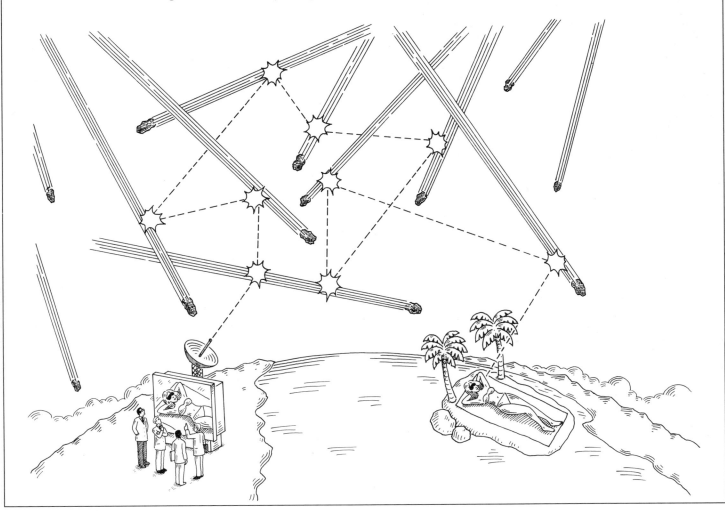

Jack Tom

MAGNETIC MONOPOLE EXHIBIT

The Erhard Wehrner Collection of Magnetic Monopoles will be on display in the following cities during May, June, July, and August:

May 5-14, London, England
May 18-26, Berlin, Germany
June 1-7, Bangalore, India
June 10-16, Alexandria, Egypt
June 19-25, Singapore
June 30-July 26, Tokyo, Japan
July 31-August 10, Santa Cruz, California, U.S.
August 14-18, Austin, Texas, U.S.
August 23-28, New York City, U.S.

Please consult local publications for exact exhibit locations, hours, and admission fees.

Paleontologist in Pinstripes?

Celebrity may soon be the link between science and the world of professional sport. America's best known major league baseball team, the New York Yankees, is planning to add Harvard paleontologist Stephen J. Gould to its player roster. Gould is perhaps the most widely known scientist whose passion for sports (at least one sport) has captured the public's imagination. The Yankees are expected to sign Gould to a 10-day contract (the major league minimum under baseball's collective bargaining agreement with its players). Observers say that this is reminiscent of an incident during the 1951 baseball season when the St. Louis Browns signed Eddie Gaedel, a 3'7" midget and amateur paleontologist, to play in one game. Gould is of considerably larger stature, both in and on the field.

Whither Marine Radioactive Waste?

After several false starts, the U.S. and Canadian governments have begun actively testing whether barrels of radioactive waste dumped in the Pacific Ocean pose a significant danger. Newton Goldschlag of the Joint Oceanic Experimentation Council is conducting a $900,000 study that will focus on the estimated 47,500 steel barrels dumped in the Gulf of Farallones (30 miles west of San Francisco Bay) between 1946 and 1970. The barrels contain waste from the Manhattan Project and from two nuclear laboratories in California, as well as some from the U.S. Navy. It is feared that some of the barrels are in danger of collapsing or corroding and leaking radioactive material. Goldschlag will break open several of the barrels, and over the next 5 years will monitor what happens as the contents spread throughout the Pacific. Goldschlag is unsure of exactly what is in the barrels but believes it may be a chemical mixture that includes plutonium, cesium, and mercury.

CHAPTER 6

NUMBERS AND THEIR ILK

PROOFREADERS' UPDATE JANUARY 1991: THE FOUR-COLOR MAP THEOREM

Joe Slavsky
Newark, New Jersey

In 1976, Kenneth Appel and Wolfgang Haken of the University of Illinois solved the Four-Color Problem.[1] Among mathematicians the proof was met with both jubilation—at the solution of a problem that had defied solution since Francis Guthrie first posed it in 1852—and dismay. Many mathematicians were troubled because the proof made unprecedented use of computer computation. It was feared that the correctness of the proof could not be checked without the aid of a computer.

In 1982 together with my fellow members of the American Society of Proofreaders, I took up the task of checking Appel and Haken's proof using only traditional means. Our work is funded by the U.S. Government's Strategic Defense Advanced Research Projects Agency. Each January, we issue a brief public statement about our progress. This is the sixth such report.

Three members of the proofreading team died during calendar 1990, reducing our total number to 117. It was decided to reduce the standard proofreading work week to 10 hours a day, 6 days a week; this measure was put into effect primarily because of cost restraints newly imposed by the funding agency.

This year, 43 person-years were expended in preparatory computation related to the semicritical open subsets of Φ_6 that satisfy the Bend Condition.[2] The preponderance of proofreading time, as in previous years, was devoted to iterative and/or recursive computation related to the introductory portions of the proof.[3]

The nature and importance of the task are at all times paramount in our thoughts. This is truly an exhaustive proof.

1. Appel K, Haken W. Every planar map is four-colorable. *Contemp Math* 1990: 98, American Mathematical Society. This 741-page document is a concise outline of Appel and Haken's original computer-aided proof.

2. See Appel and Haken, p. 216 ff. This is the sort of task that we (and presumably Appel) would ordinarily delegate to troublesome graduate students. However, the overarching importance of this proof precludes us from delegating any portion of it, no matter how straightforward, to unproven personnel.

3. Outlined in our annual statement, issued in January 1983.

These figures illustrate what the four-color map theorem means. To color a planar map, one needs no more than four colors to ensure that no touching countries have the same color.

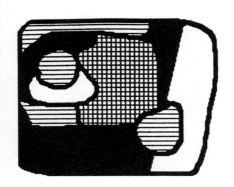 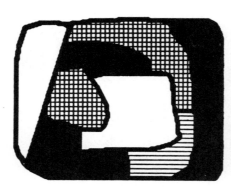 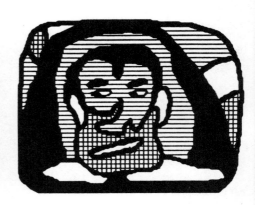

AN EXPERIMENTAL DETERMINATION OF THE NATURAL BASE OF LOGARITHMS

Gorden Videen and William S. Bickel

University of Arizona, Tucson, Arizona

Abstract

The method used to determine the fundamental constant π has been adopted to determine the natural base of logarithms, e. The accuracy is expected to be as good. The precision, however, has been found to depend on the font.

Introduction

The experimental determination of the fundamental constant was discovered at least 40,000 years ago by the Egyptians.[1] So many other cultures have used this method (the Babylonians and Chinese, for example) that its history probably extends even farther into the past. The method is quite simple and is derived from the definition of π. It consists of drawing a circle and its diameter, d (see Figure 1), and determining the number of times that the diameter can be wrapped around the circle. Such a procedure can be carried out by using a pencil, paper, and some string.

With the advent of high-speed digital processing, we can facilitate the measurement through the use of a computer program. Such a program draws a circle on a high-resolution monitor having a diameter of n pixels. The ratio of the number of pixels used in drawing the circumference to the number of pixels used in drawing the diameter is the value, π. Such measurements have been routine since the advent of computers. Shanks and Wrench performed such a calculation of π to 100,000 decimals in 1962.[2] We are not surprised that the paper in which they report their result fails to include the resolution of the monitor used. Such a high-resolution monitor is undoubtedly classified.

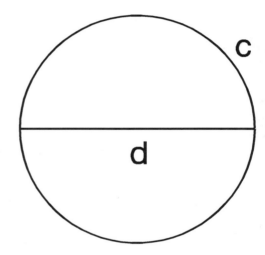

Figure 1. The method used in determining the value π. After measurement of the diameter, d, and the circumference, c, π is the ratio of $c{:}d$.

The Experiment

Figure 2 illustrates the experimental determination of the natural base of logarithms. The method resembles closely the method used to determine the value of π. First, the length of the line segment b is determined. Second, the length of the arc a is determined. The ratio of these quantities $a{:}b$ is the value of e. Preliminary analyses with a meter stick and French curves revealed results for e that were well within experimental error.

Without a doubt the greatest difficulty in measuring the value of the natural base of logarithms is in obtaining an adequate e. To reduce measurement error, it is nec-

Figure 2. The method used in determining the value *e*. After measurement of the segment *b*, and the arc length *a*, *e* is found to be the ratio of *a:b*.

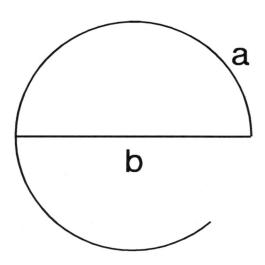

essary that the *e* be as large as possible. Enlargement of normal-sized fonts is possible by making a photomicrograph, such as the one shown in Figure 3. From this photomicrograph we can see the difficulties in accurately measuring *e*. The graininess of the image, due to the printing process, is accentuated in the magnified image, and the resulting fuzziness effectively hampers an accurate determination of *e*. The development of multiple font sizes in word-processing programs greatly simplified one task, whereas before this development an accurate determination was a virtual impossibility.[3]

Encouraged by preliminary results, we set out to do a computer analysis of the problem. As outlined in the introduction, we wrote an assembly language program to measure the quantities *a* and *b*, by counting the number of pixels used in forming these arc and line segments on our computer monitor. This was performed for a variety of different fonts. The results are shown in Figure 4.

Results and Discussion

The values determined for the natural base of logarithms using the Swiss and Dutch fonts are well within the experimental error of the actual value (*e* ≈ 2.720 ± 0.002). However, for the Greek font the value determined for *e* is many standard deviations from the accepted value. Repeated measurements revealed similar results. From these measurements we conclude either that the accepted value of *e* is incorrect, or that there

is a fundamental problem with our methodology. Even though the value of *e* is determined by well-established methods of calculus, it was difficult to believe that some aspects of the experiment were faulty. However, after reviewing the procedure, we came to the conclusion that the Greek fonts supplied with our word-processing software were faulty and the source of the problem. After many conversations with the customer support office[4] we decided to purchase a different word-processing program. We were surprised to find that this new program showed the same phenomena. Both Swiss and Dutch fonts produced values well within experimental error, while the Greek font produced values that were wholly inaccurate.

From these results, only one conclusion can be reached: The ancient Greeks, with their ill-equipped alphabet, were incapable of determining an accurate value for *e*. Thus they were unable to develop the calculus. The actual date of origin of the calculus is debatable. Traditionally, people favor the 17th century, giving credit to Newton and Leibniz. Some feel that its origins lie with Kepler in his calculations of areas of portions of orbits. Still others favor the ancient Greeks Archimedes and Democritus, who calculated various volumes. Since *e* is fundamental to the calculus, and the

Figure 3. Photomicrograph of *e*.

Character	Style	Value determined
e	Swiss	2.720±0.002
e	Dutch	2.71±0.01
ϵ	Greek	3.983±0.003
ϕ	Greek	2.343±0.001

Figure 4. Illustrations of different fonts and the values determined from them.

alphabet used by the ancient Greeks could not be used to determine its value accurately, the priority of discovery of the calculus must therefore lie with the later Europeans.

Finally, in our research we have discovered a truly remarkable event. The Greek character φ bears the same characteristics of the constants π and e, and the value was determined by using the very same experimental method to be 2.343 ± 0.001. Since φ is not yet known as a fundamental constant and no other fundamental mathematical constants have this value (for instance, Euler's constant is $\gamma \approx 0.57721$), we can only assume that we have discovered a constant that is the cornerstone of a yet undiscovered branch of mathematics. Perhaps with this value known we will soon be able to discover this branch.

Conclusion

This paper shows the importance of the oft-overlooked field of experimental mathematics and the role it plays in making new discoveries. We used a well-established method to determine a known fundamental constant, e. We then extended this method to find the value of a currently unknown constant, the importance of which can only be postulated.

LETTERS TO THE EDITOR

Integrity

To the Editor:
I am on the editorial board of ***JIR*** for the areas of psychiatry and neurology, having trained in both as a result of a chronically indecisive temperament. From this perspective, may I note that most of us in medicine consider the ***JIR***, satire journal or not, to be the best source of reliable medical research data. At this point, the ***JIR*** is the only journal publishing medical studies which has never been accused of publishing faked data or data of any sort, for that matter. I think part of its success in this area is attributable to the ***JIR***'s method of peer review: Due to an early misunderstanding about the definition of "peer," its articles have traditionally been reviewed not by equals (the first Webster definition) but by titled nobility (the second definition) who have no knowledge of the field involved but who dress very well and have good manners. I strongly support this approach.

Robert S. Hoffman, M.D.
Peninsula Neurology Associates
Daly City, California

1. Beckmann P. *A History of π (pi).* New York: St. Martin's Press, 1971.

2. Shanks D, Wrench JW Jr. Calculation of π to 100,000 decimals. *Mathemat Comput* 1962; *16:*76–99.

3. It should be noted that characters may also be enlarged by using the magnification mode on a copier. A quantative determination of e using this method is inherently flawed, since the lenses of copy machines are notoriously plagued with distortion, leading to erroneous values.

4. Customer support representatives invariably tried to convince us of the infallibility of their product in this matter and suggested that we purchase the updated version.

SIMPLIFIED MATHEMATICS

Ben Ruekberg
Cranston, Rhode Island

As John Allen Paulos points out in his book *Innumeracy: Mathematical Illiteracy & Its Consequences*, "[some] of the blame [for the inability of students to do math, aside from 'extreme intellectual lethargy' (p. 89)] . . . must ultimately lie with teachers who aren't sufficiently capable . . . [themselves]" (p. 75). He suggests that others, who *are* capable in mathematics, help instruct. Toward this noble end, I submit a convenient and useful shortcut in dealing with fractions, which are the bane of young students.

While teachers often instruct students in simplifying fractions, they usually restrict their instruction in this method to cancelling zeros. An example of this is:

$$10/100 = 1\underline{0}/1\underline{0}0 = 1/10$$

What they fail to point out is the much more useful expedient of cancelling non-zero digits or groups of digits that lie adjacent to the slash. Since this method may be novel to many readers, and because of possible confusion between the slash indicating "division" and the slash indicating "crossing out" (and also because it is a great way to save space), the cancelled digits will be underlined.

For the first example, consider the fraction 16/64. A "6" lies on each side of the slash, and, thus, can be cancelled:

$$1\underline{6}/\underline{6}4 = 1/4$$

Another example of one-digit simplification of fractions is 19/95:

$$1\underline{9}/\underline{9}5 = 1/5$$

The simplification can also be done with two-digit factors which lie in the same order on either side of the slash. In 133/3325, the 33s cancel:

$$1\underline{33}/\underline{33}25 = 1/25$$

The method also works well with larger cancellations, as illustrated by a few random examples:

$$30\underline{405}/\underline{40540} = 30/40 \text{ (by conventional cancellation, 3/4)}$$
$$1\underline{501}/\underline{501}334 = 1/334$$
$$47\underline{641}/\underline{641}633 = 47/633$$
$$467\underline{956}/\underline{956}955 = 467/955$$
$$32\underline{727}/\underline{272}725 = 3/25$$
$$129\underline{612}/\underline{961}289 = 12/89$$
$$2\underline{77777}/\underline{77777}56 = 2/56 = 1/28$$
$$4\underline{848484}/\underline{848484}70 = 4/70 = 2/35$$
$$7\underline{407407}/\underline{407407}385 = 7/385$$
$$4\underline{9999999999}/\underline{9999999999}8 = 4/8 = 1/2$$

This method works in simplifying improper fractions as well:

$$493\underline{991}/\underline{991}99 = 493/99$$
$$4\underline{32}4/\underline{32}43 = 4/3$$
$$56\underline{504}/\underline{504}5 = 56/5$$
$$67\underline{402}/\underline{402}4 = 67/4$$
$$23\underline{828}/\underline{828}8 = 23/8$$

The skeptical reader might believe that I have carefully picked cases that work rather than picking numbers at random. I will demonstrate otherwise, with the cancellation of two consecutive numbers and multiple examples using the second, all with three-digit numbers cancelling; then with two fractions which simplify to the same fraction, with four-digit numbers cancelling.

702 (followed by 7): $18\underline{2702}/\underline{7027} = 182/7$
703 (followed by 7): $13\underline{3703}/\underline{7037} = 133/7$
703 (followed by 57): $3\underline{703}/\underline{70}357 = 3/57$

Two fractions which simplify to the same value:

$$123\underline{762}/\underline{37623}648 = 12/3648 = 1/304$$
$$103\underline{135}/\underline{31353}040 = 10/3040 = 1/304$$

Only when our school systems adapt innovative approaches to handling mathematical manipulations, such as this, can America hope to continue to compete in today's world. We cannot afford to delay much longer.

REFERENCES

Paulos, J. A. *Innumeracy: Mathematical Illiteracy & Its Consequences.* New York: Hill & Wang, 1989.

THE TOURING MACHINE

Douglas Loss
Williamsport, Pennsylvania

I must confess I never expected to be asked to do a job like this. Probably everyone who has ever written reviews for computer magazines has complained about the mind-numbing parade of products that are almost indistinguishable from one another. I know I did. If that's why the editor gave me this assignment, he has cured me of that line of complaint. This product is definitely distinctive.

The Touring Machine is Imitation Intelligence, Inc.'s (I[3]) first commercial product. If I had to summarize this whole review in one sentence, I'd say that this computer owes nothing to any computer ever made before.

System

The Touring Machine consists of a keyboard, monitor, and processor. Optional peripherals include a printer, single and dual floppy disk drives, a hard disk drive, and a job control system of sorts. Bundled with the basic system are three pieces of software: a word processor, a spreadsheet, and a database manager. I'll try to give my impressions of them all.

The first thing that you notice when you set up the Touring Machine is the unique keyboard (Figure 1). The key layout isn't the industry semistandard QWERTY or even the widely touted Dvorak. The keyboard has one row of 128 keys. It took me a minute to recognize the sequence: from left to right, it was ASCII! As the I[3] literature says, "This is the first keyboard designed expressly for personal computer use, using the industry standard ASCII code." I doubt whether I could describe the keyboard any better than they do: "A series of control and communication keys, some random punctuation, the numerals from 0 to 9, more random punctuation, the capital letters from A to Z, more random punctuation, the small letters from a to z, and yet more random punctuation."

At first sight, the monitor appears to be a standard amber-screen display. Closer inspection shows a one-quart kerosene tank mounted high on the back of the unit. The screen itself appears to have three layers: a glass outer layer, a catalytic igniter grid, and a fine network of tubing with tiny jets at the intersections of the tubes. Presumably, this is what is meant by I[3] reference to "triple-tracked window" and not the currently-in-vogue screen formatting techniques.

I can't verify the claim of "inherently radiation-free," but the KJD (kerosene jet display) technology does give a pleasant flicker to the amber characters. Display updates aren't as fast as we have come to expect from CRT displays, but considering the amount of plumbing that must be needed in a display of this sort, they are remarkably fast.

I[3] promised to have a full-color KJD terminal available next year, using injected chemical salts to provide the colors.

The whole unit is steam-powered. (I admit I don't know how it's done, but there are no electrical cords

Figure 1

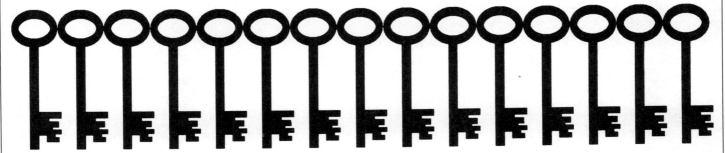

on the unit.) A notice on the door of the burner compartment mentions that for best results you should use apple wood. This might be a subtle dig at a competitor, since it doesn't appear anywhere else on the machine or in the documentation. The burner heats a 3-liter

Figure 2

buffer that uses "time-proven hookah technology with optional mint injection" to power a chip set with a unique substrate of arsenic hydride rather than the more common silicon or gallium arsenide. This AsH3RAM is arranged in a "multitier omnibus with a trunk terminator" (Figure 2).

As for the memory map, I again quote from the documentation: "A paging interpreter controls a mapped, token-passing queue, directing data into the 76-seat storage area (76 seats not including the hardware driver; Figure 3). The driver is normally outside the storage area doing collision detection but is paged as necessary by the interpreter to take control of the CPU through the custom right-handed operating system." I have no idea what this means.

The CPU is, of course, a custom design. It passes all data through a binary logic utility register (BLUR). Then it's on to dual data lines, each with its own cache register, for minimal wait states. I³ has added some interesting new op-codes to the design, too. I don't have a good feel for their best uses yet, but they sound as though they should be useful. They are DWIM (Do What I Mean), BLFDN (Blink Lights Furiously and Do Nothing), and CMFM (Come From). There are a few more that ought to be quite useful in AI research; they are PLZ (Please), JFW (Jump If Willing), and WGOIT (What's Going On In There).

Peripherals

I³ peripherals show the same approach to the computer field as the main system. The McCullough chain printer

has a special high-traction paper feed, and our review model was equipped with the optional studded platen designed "for heavy duty off-line printing." The printer is rated at 7,600 characters per gallon and has a plug lifetime of 3,000,000 characters. Frankly, its print quality is nothing special, but its noise level is. The optional ear protectors, modeled on shooter's ear protectors, should be standard equipment.

The printer is designed to run on a mixture of unleaded gasoline and two-cycle oil. We found that getting the mixture ratio wrong could cause the engine to miss, usually losing charcters and occasionally printing some surprising things that could be indicative of the use of the AI op-codes I mentioned above.

I³ Sunbeam top-loading single disk drive (Figure 4) seems to perform just as we have come to expect a disk drive to. The lack of a door or latch over the drive slot was disconcerting, especially since the drive was in the habit of flipping the disks high into the air whenever the Touring Machine decided that it was done with them. After a few days, I began to be able to guess with some success when a disk was due to come flying out, and I became pretty good at catching them before they flew behind the radiator.

Figure 3

The dual disk drive is of a totally different design. The drive is topless, allowing you to see the surging disks in all their detail. It requires double-sided, DD disks, and is quite fast, with an information transfer rate of 5 megabawd (that's from the documentation, and may be a typo).

The Black & Decker Winchester hard disk attachment is labeled as "dual speed reversible." I suppose dual speed means high-speed tracking to the proper sector, then normal speed information access. Reversible doesn't suggest much of anything to me. Neither does the documentation, which is very scanty at this point. It does mention that the standard unit has a 1/4 bit rate, but that a 3/8 bit rate is available for those with large data transfer needs.

Figure 4

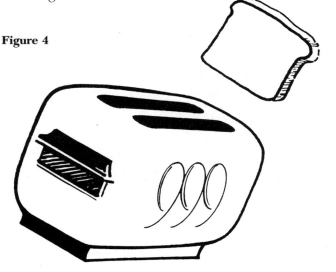

The Smith & Wesson job control output device (Figure 5) is a peripheral that has little in common with anything else in the computer world. As a device to control the physical environment, it has some strong points, such as not needing any special interfaces to transmit its controlling impulses to the desired receiver. The version I tried was the .38 ACP model. After an initial zeroing-in installation procedure the device consistently put six controlling impulses in a 4-inch group at 40 yards in 15 sec, quite adequate for its intended uses. A 9-mm Parabellum version is also available for the European market.

Software

The software included with the system is pretty much what you might expect from I^3 from reading this review.

The spreadsheet, Cut & Run, has been specifically designed for tax advisors. I suppose that's the reason that two undated one-way airline tickets to Barbados are included in each package. I'm no accountant, but Cut & Run seemed to provide all the functions that are ordinarily expected in a spreadsheet program. Its major distinguishing feature was the dual report mode. In this mode the first report printed is faithful to the numbers

input, while the second report shows widely lower profits and sometimes even losses. Since these reports aren't identified differently in any way, it is advisable to keep careful track of which is which. There is an option available that might help, in which the printer smudges certain groups of characters to simulate tear stains.

The word processor, Search & Destroy, handles the standard text editing features adequately. I do wish the machine gun noises and screams that you get when you delete any text at all, low volume though they are, were toggleable. Also, the program's predilection for dropping into "scorched text mode" at the entry of any incorrect command sequence and deleting chunks of text (with guns and screams) seemed less than enthralling. And when I tried to use the word count function the first time, I got a figure that was at least double what I had expected. I tried it again and got a figure that was about 20% higher still. When I tried it a third time, the command line displayed this prompt: "How many would you like there to be?"

The previously announced spelling checker, Slash & Burn, wasn't a part of the software bundled with the Touring Machine. Instead, Search & Destroy has an "Englishification mode," which is supposed to serve the same purpose. When I used it on a short document with a few purposely misspelled words, it corrected about half of them. Unfortunately, it changed about the same number of correctly spelled words into misspellings.

Figure 5

The database manager, Dominatrix, is the star of the I^3 software line. In the short amount of time that I had to try it, I could tell that it was very powerful but quite demanding. If you make a mistake, it forces you out of the workspace and ties you to a complex series of keystrokes before it allows you back in. Record formatting is free-form within strictly defined limits, but those limits are steel-tight. On the occasions when I tried to "bend" the formatting, Dominatrix forced me to type, "I was wrong to ignore the requirements of the program. I have learned a lesson and will obey the program commands from here on," 50 times before it would let me access any of my other files. The packaging of the documentation is imposing, too. The manual is tightly bound in black leather and includes a strict and demeaning tutorial on the program's use.

The program requires the McCullough chain printer with the studded platen. The Smith & Wesson job control output device is strongly recommended, but frankly, I was afraid to connect it.

Support

The I[3] warranty (Figure 6) is not really any different from most others in the computer industry, but it is refreshing in that it doesn't attempt to hide its guarantees or lack thereof in legalistic phrases. It's nice to know exactly where you stand.

Both for the purposes of this review and for my own curiosity, I decided to contact the support section of I[3]. There was no support phone number listed anywhere in the Touring Machine package, but I managed to get one by checking around and calling in a few favors. When I called, a woman answered. After an initial attempt on her part to deny that I'd reached I[3] at all (I wasn't buying any of it), I was connected with the support section.

A man answered with a heavy but indeterminate foreign accent. I explained my problems to him, and he said, "Wow, that's a tough one. Tell you what, give me your name and phone number, and I'll get back to you as soon as I have an answer." I said that I couldn't give him a number but I'd appreciate any advice off the top of his head. He responded, "OK, I've got it. I'll call you as soon as I figure it out."

I started to say something, but he hung up. After a little reflection I decided that this man didn't actually understand English. He listened to what was said, then recited the first line, which he must have learned phonetically. The caller was then supposed to give a name and number, and he said his second line, also phonetically, and hung up. That was as far as I got with I[3] support.

Figure 6

OUR OWN CLEAN DEATH WARRANTY

Full Operation Is Absolutely Guaranteed to Be a Figment of Your Imagination.

No utility, nay, no usefulness at all is promised, implied or will be delivered to any purchaser of these items. Any problems, errors, or liabilties incurred due to the use of our products must be your fault. They will be disavowed, ridiculed, and held cause for retribution by the legal arm of I[3] in grand congress assembled.

Failure to sign this agreement will earn you a punch in the nose, buster. I mean, where the hell do you get off even reading this far down in a warranty that you'll never be able to hold us to, anyway. We have operated under a lot of names in the past, and new names are a dime a dozen, fella.

Viewed as separate parts, each component of the Touring Machine seems very unusual, but taken together, everything seems to fit in a demented way. Maybe this system would be best for someone who is new to personal computers. As for me, I'm too tied to traditional systems, boringly similar though they may be.

THE SIGNIFICANCE OF THE NUMBER 17 TO MATHEMATICIANS OF MY GENERATION

Lee F. Mondshein
MIT, Cambridge, Massachusetts

I would like to discuss a subtle social practice, largely unuttered for want of a name, which I shall call "significant digits for significant others."[1] A little personal history will clarify this pervasive phenomenon. When I was studying mathematics, several of my professors, searching for a "random" number to illustrate a point, would predictably say, "Pick a random number, like 17." After observing this evidently nonrandom behavior for several months, several of us began to wonder whether we were victims of some obscure conspiracy or witless witnesses of some advanced number-theoretic joke.

As I recollect, we were told—upon being so bold as to enquire—that one professor at Princeton had become notorious for the choice of this number, which had thereby achieved a special, affectionate character of its own. His students, in light-hearted tribute to their master, had acquired the same quirk. I am unsure how many generations of mathematicians these "significant digits" have mastered in this same manner, but I am curious to know whether any reader, near or distant, has experienced their appearance.

What intrigues me is the fact that mere digits can attain "significance" to such a degree that their repetition becomes a kind of recursive mental gene, regenerating itself as a result of contact with the bodies (or, rather, the minds and mouths) of a special society devoted (or commandeered) to its perpetuation. Have other *JIR* readers experienced this recursive, random 17 in the course of their training? I wonder what a genealogy of the 17ers would look like. What is the identity of the academic don who played the role of Adam to this human subspecies?

Clearly, other such "digital ideo-genes" exist—for instance, the googolplex, which was widely spread in its first generation through contact with the book in which it was initially used.[2] Each of these "genes" spawns a whole family of people who are linked to each other not only by textual or oral familiarity or habit, but through shared intellectual interests and linguistic preferences that amount to making them "significant others" in a sense.

This wild spread of harmless, shared proclivities is a reminder that language is not only a channel of information and control, but also a lively medium for uncontrollable generation and transformation, however small or great.

Are there any physicists, chemists, astronomers, biologists, geologists, engineers, meteorologists, economists, psychologists, dentists, doctors, librarians, or neuropharmacologists out there who are willing to reveal and explain their special significant numbers to the larger scientific community?

---

1. "Significance" is a term with many meanings, arising in many contexts, from academic discourse to common parlance. Two such contexts are the computational and the conjugal: Many readers are acquainted with "significant digits" and the issues of computational precision underlying this notion, and many are familiar with the emergence of "significant others" and the issues of conversational imprecision engendering that concept.

2. Interestingly, some complementary numbers seem to generate correspondingly complementary groups; many of the devotees of 007 and 770 are probably unaware of each other's existence.

SWEET SEVENTEEN

Dudley Herschbach, Nobel Laureate (Chemistry, 1986)

Harvard University, Cambridge, Massachusetts

Lee Mondshein has described "The Significance of the Number 17 to Mathematicians of My Generation," (*JIR* 1991; 36:2) observing a strong propensity for mathematics professors to pick 17 when searching for a "random" number to illustrate a point. He attributes this to "a kind of recursive mental gene," by which generations of professors propagate the whimsy of an ancestral teacher. Mondshein wonders who might have had "the role of Adam." This prompts me to report evidence on behalf of two candidates.

The late Professor John Van Vleck, a distinguished Harvard theoretical physicist (Nobel Laureate, 1977), also invariably used 17 whenever he had occasion to refer to an arbitrary number. Curious about that, I asked him one day after class why he always chose 17. He told me a story about the great mathematician Karl Friedrich Gauss. As a young man, Gauss solved a classic prob-

lem that had stumped the Greeks, by showing that regular polygons with $2^{(2n+1)}$ sides could be constructed by "ruler and compass." The ancients knew about the easy cases $n = 0$ (triangle), and $n = 1$ (pentagon), among others. In 1801, Gauss was elated to discover how to do the $n = 2$ case (septendecagon, a 17-gon) and likewise $n = 3$ (a 257-gon), $n = 4$ (a 65537-gon), etc. This discovery was the key to a far-reaching theorem dealing with algebraic numbers and the solutions of polynomial equations. It also confirmed Gauss's decision to be a mathematician. For Van Vleck, and surely for many others, Gauss was the septendecal Adam.

I recently learned of another candidate. Oona Ceder, a 1989 graduate of Harvard, stopped by and told me of her summer adventures living with a group of friends "Indian-style" in northern Maine. She described the large conical tepees that they had constructed, following a traditional tribal pattern. These tepees have 17 poles located at the vertices of a regular 17-gon! Long before Gauss, such septendecagonal tepees may have been designed by a Native American Adam or Eve.

MESSAGES FOUND ENCODED IN THE DIGITS OF π

Tim Bell and Rod Harries
Christchurch, New Zealand

The number π (≈3.14159265358979323846) is currently known to many thousands of decimal places, and there is every indication that many more digits remain to be computed. Since only the first 30 digits are required to compute the circumference of the known universe to within a millimeter of accuracy, one wonders what the purpose of all the others might be. It is our hypothesis that these digits contain coded messages.

Although π has an infinite number of digits, it is not yet known whether every possible combination of digits eventually occurs in the sequence. If this turns out to be the case, then a coded form of every piece of literature could be found if you looked far enough. Even all the works of Shakespeare would be found—even every alternative that Shakespeare considered but didn't write down.

Since all digits past the thirtieth place are of little use to the geometer, we might not have to look infinitely far for a coded message. It might appear near the beginning. On the basis of this observation we fed the first 100,000 digits of π into a computer and then had them analyzed for correspondence to portions of Shakespeare. To our surprise, the computer very rapidly located many familiar fragments from the Bard's writings; specifically, the page numbers.

It is interesting to note that around the time of Shakespeare's death, π was known only to about 35 decimal places (reported by Snell in 1621), so it is all the more remarkable that Shakespeare's work contains parts of π that he could not have known about.

Many colleagues did not seem to share our excitement over the discovery of Shakespeare's page numbers, so we then considered more complex codings that might hide messages in the digits. We reasoned that if some intelligence had placed a message in a numeric form, there must be a correspondence between the 10 digits and the letters of the alphabet. After some experimentation, the code in Table 1 was found to reveal messages of the nature that we were looking for.

Table 1. Relationship of 10 Digits to Alphabet.

Number	Letter	Number	Letter
0	k	5	t
1	s	6	e
2	p	7	i
3	o	8	u
4	space	9	h

Decoding the digits of π using this system, we encountered a portion of the Bible around the 180th digit (. . . 59644622948954930381 . . .), which decodes as "the epph uht hokous." The first word of this phrase is a direct quotation from Genesis, Chapter I (Authorized Version): "In *the* beginning . . ." This is not the only place in which the fragment appears in the Scriptures; a concordance shows that "the" appears frequently in many books of the Bible, adding weight to the significance of its appearance in π. What makes our find all the more remarkable is that the writers of the Bible regarded π as 3.[1] This is of course because 3 is a natural number, and, as Dedkind observed, "God created the natural numbers, all else is the work of man."[2]

We should not disguise the fact that this interpretation has been challenged by some of our colleagues.

They point out that those familiar with Linear B[3] and with the highly elliptical and contracted forms of Greek found on icons[4] will be inclined to regard the phrase "the epph uht hokous" as an abbreviation of "Θοζ επφαιυει ΄υποχορεται ΄ο κοινονουζ," to wit, "God becomes known (to and) withdraws (himself from) the sharer." The relative merits of these interpretations are clearly fertile areas for further research.[5]

More skepticism from colleagues prompted us to look for further evidence of codes in irrational numbers. The number e (2.71828 . . .) seemed like a good candidate, and at the 95th digit we found the sequence . . . 166427 . . . , which translates to "see pi." This has prompted us to intensify our investigation of π, and further messages are expected to be found as soon as is forthcoming.

---

1. I *Kings* 7:23.

2. *Numbers* 1:19.

3. Chadwick J. *The Decipherment of Linear B*. Pelican, 1961.

4. *Inside Macintosh*, Vol. 1. Reading, Mass: Addison-Wesley, 1985, p 32.

5. *Psalms* 1:3.

To The Point

Sir:
The chart on page 41 of your November issue (*JIR* 1990; 35:5) contained an error. The value in column 4, row 16 was listed as 392411115.00000000 000118. It should be 392411115.00000000000117.

K. Bloom, Ph.D
The Shaw Institute
Phoenix, Arizona

Counterpoint

Dear *JIR* Editor:
There was an error, probably typographic, in the chart on page 41 of the November issue. The number in column 4, row 16 should be 450, not 392411 115.00000000000118.

Megan Lester, Ph.D., M.D.
Skrock Associates, Ltd.
New York City, New York

Beside the Point

Dear *JIR*:
Some of your readers might not be aware that the November issue presented a most interesting number.

The chart on page 41 (see column 4, row 16) contained the value 392411115.00000000000118. This is a most fascinating number! This single number contains: (a) the value of Plankk's [sic] constant; (b) the average distance in meters between the moon and the earth; (c) the characteristic frequency of the distress call of the worker ant of the species *Aphaenogaster rudis* when it is deprived of *Sanguinaria canadensis,* its favorite food; and (d) the molecular weight of silicon.

And to think that some people wonder what there is to love about science!

Pelham Grenville
Wodehouse, New Zealand

THE BINARY ABACUS

Gary Garb
Bensalem, Pennsylvania

The instrument shown in Figure 1 has been called the "binary abacus" by its discoverers.[1] It was found in the archives of an ancient Buddhist temple near Kyoto, Japan, in 1837. With it were extensive notes and treatises on mathematics and on the use of the device by 12th century mathematician and philosopher Mitsuyana.

The binary abacus permits much faster arithmetic calculations and is much easier to learn and use than traditional models. Numbers are represented by the positions of the beads (up or down) on each wire in the frame. The arithmetic is performed by changing the position of the beads according to a very simple set of rules.

Going from right to left, the beads represent successive powers of 2 in the same way that we write numbers as a string of digits representing powers of 10. The first bead on the right, called the "1-bead," represents the value 1, which is 2 to the 0 power (2^0). The second bead from the right (just to the left of the 1-bead) is the 2-bead and represents the value 2 (2^1). The 4-bead is third from the right and represents the value 4 (2^2). The remaining beads, in order from right to left, represent successive powers of two: 8 (2^3), 16 (2^4), 32 (2^5), and so on.

A bead is raised on the wire to indicate the presence of the value it represents. For example, if the 4-bead is up, then the value in the frame is 4. To represent a value that is not a power of 2, combinations of beads are raised to make up the value.

The magnitude of the values calculated is limited only by the size of the device. The 13th century Emperor Shamuramatsa was told a Buddhist concept by the Zen monk Mishugi: Every day, a bird with a silk scarf in its mouth flies over the greatest mountain on Earth and brushes it with the scarf. The length of time required for this to reduce the mountain to dust is but one moment in the life of the Buddha.

The Emperor wanted to immortalize himself by calculating that time in days. Consequently, he constructed a binary abacus with 200 beads and capable of holding a value of $2^{200} - 1$, or approximately 10^{60}. Three years into his calculations, the great earthquake of 1299 reinitialized the device. In a rage the Emperor ordered all binary abaci destroyed and exiled Mishugi to China, where he later met Marco Polo.[2]

The origin of the binary abacus is unknown, but clearly it must have sprung from binary counting. The development of the decimal system and decimal abaci followed logically from finger counting. The vigesimal (base 20) number system used by the Mayans undoubtedly resulted from counting on fingers and toes.[3] Following this reasoning, there has been speculation that the originators of the binary abacus restricted counting to only their thumbs or other paired body parts.[4]

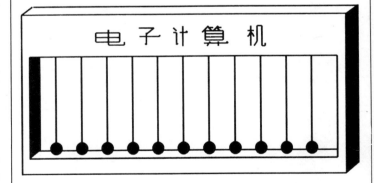

Figure 1. The binary abacus (c. 1030 AD).

1. Mitsiya T. *Early Eastern Asian Computing Devices*. Tokyo: Kinahara Publishing Co, Ltd; 1957, in Japanese.

2. Most of Mishugi's work was not utilized during his lifetime. In recent years, Benoit Mandelbrodt has expanded upon other of Mishugi's ideas in developing the field of fractal geometry.

3. There is evidence of an early Central American native people who practiced total nudity. Not surprisingly, they developed a base-21 and base-22 number system (unpublished report by the 19th century French explorer Comte Penilly).

4. Mitsiya, *op cit.*

Styles, trends, and tidbits culled from leading research journals

by Alice Shirell Kaswell

The November 1990 issue of the research journal **Vogue** presents an impressive array of findings. A report on p. 19 describes totally new technology for the eye area. The technology is not a conventional creme, but a multiphase liquicreme so unique it has a patent pending. The principal experimenter, Estée Lauder, reports that it literally creates a new surface on the skin with ultrahydrating emollients. A report on p. 20 describes Lancôme's maquiéclat, an amplifier for skin beauty. A contrasting report (p. 43) describes a StayRich formula of microtextured pigments and 60% emollients. On p. 73, investigator Elizabeth Arden presents another in a continuing series of reports on a series of experiments involving ceramide time complex capsules. Each capsule is unique.

Recursive Perfume

Jean Patou presents findings (p. 81) that his 78-year-old grandmother is wearing fragrance and living with her 28-year-old fencing instructor. **Vogue** also has a related report (p. 86) that involves a rejuvenating *système biolage* wave. A logic treatise on p. 107 gives fresh insights concerning a perfume that reminds you of a woman who reminds you of a perfume. A chemical

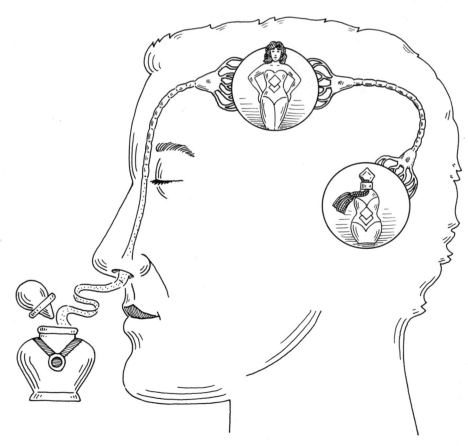

investigation on p. 108 describes Oil of Olay Sensitive Skin Creme's principal components (rich, rich, rich emolients) and characteristics (pure richness, pure luxury, pure sensitivity, pure creaminess, and pure silkiness).

Puzzling Pants and Floor

Visual Eyes is a whole new kind of color. It is also a whole new range of color. Details are presented on p. 156. *Vogue* investigator Carol Milbank's multidisciplinary research paper (pp. 121–80) eludicates physical phenomena that include skinny stovepipe pants and a Valentino floor that turns into a public dressing room. An abstract on p. 195 reveals that lotions sometimes call it a day.

Fahrenheit Eau de Toilette

Vogue also has a brief report on p. 236 about the case of a psychological fanatic who craves Vitabath. A puzzling thermodynamics investigation by Christian Dior (p. 239) analyzes eau de toilette on the Fahrenheit scale. Senior investigator Paula Kent Meehan, Redken Laboratories' founder, chairperson, and CEO, describes her laboratory's extensive research into the scientific analysis of hair (pp. 240–242) and presents a convincing case for an intelligent solution for beautiful, healthy hair. A corollary, too, is presented: an intelligent solution for hair fashion freedom. Redken's CAT protein network system beats styling burnout and static with super-conditioning. Moreover, it replenishes heat-harassed hair. (Free samples, *Vogue* reports, are available by calling 1-800-382-4100.)

Intelligent Wool

Vogue also describes (p. 249) an ongoing experiment involving experienced language instructors who lecture to a semicircle of 15 pedalling pupils. A topologic analysis by mathematician Ralph Lauren (p. 253) concerns certain aspects of a world without boundaries. An interdisciplinary research project (outlined on p. 269) is working with pure virgin wools that say and do smart things.

Dress Ex Machina

Vogue investigator Jenny Capitain ponders (pp. 332–337) the contrast between a machine that works to uncover the body's secrets and a bare lace dress that still leaves a few secrets to the imagination.

The Engineered Bosom

Finally, *Vogue* presents an unusual taxonomy paper (pp. 338–344), in which investigator Polly Mellen discusses the prominent placement, in the Gottex collection, of the engineered bosom.

Clarifying Orange

To look healthy and trim over its bones, skin needs a step beyond cleansing, according to the research journal the ***New York Times***. A report (section 6, p. 7) in the January 20, 1991 issue explains that the hitherto missing step is exfoliation. There is evidence indicating that, for the process to happen fully, Clinique Scruffing Lotion is required. A report on p. 13 of the same journal outlines the mechanism whereby pore-tightening Melilot, clarifying orange and lemon, and many other natural experts unearth makeup. An abstract on p. 38 concerns a sporadically reported phenomenon: The sun shines at least 363 days a year.

Ladies' Effortless Power

The January 1991 issue of ***Golf*** (vol. 33, no. 1) presents investigator Johnny Myers's findings (pp. 35–40) on professionals who form an "L" at key points in their swings. Myers concludes that one should take a club and freeze oneself. A metaphysics essay on p. 54 explains that researchers at Rolex have developed a perpetual day-date. A report on p. 120 presents FTM's encouraging results with a progressive cavity. ***Golf***'s research staff presents an exhaustive review (pp. 155–162) of its proprietary techniques, including a method for eliminating the possibility of a "blow-up" hole and a discussion of guidelines for touching leaves, pine cones, and other natural objects. This issue of ***Golf*** also contains (p. 34) a summary of significant recent research findings, including sand magic and ladies' effortless power.

Jack Tom

Squinting and Puckering; Little Chicken Legs

Vogue investigator Shirley Lord (pp. 198–221) analyzes results from a number of research facilities, among them cosmetics giant Kanebo (which apparently has proved that appropriate sounds can stimulate skin at beauty salons) and Collagen Biomedical (whose research in- volves the lines that come from years of puckering and squinting through smoke rings). Lord concludes that Ricci-Club, from Parfums Nina Ricci, mixes citrus with oak moss and patchouli. Lord also finds that Jean-Paul Goude had little chicken legs, and that dancing was good for his body, and that he liked it.

CHAPTER 7

HUMAN NATURE IN THE WILD

TITULAR DOMINANCE IN *I LOVE LUCY*

Wilson Kosmowski, Ph.D.
Hong Kong

Abstract

The relative societal prominence accorded to male and female American marriage partners is contrasted for the period 1951 through 1957. Relative esteem is derived and quantified from data contained in historic records.

Method

During the 1950s, many Americans observed and kept detailed notes of their neighbors' activities. The records concerning two such families were examined. One family, the Ricardos, consisted of a husband, Ricky, and a wife, Lucy. In 1953 they became the parents of a son, little Ricky; the second family, the Mertzes, consisted of a husband, Fred, and a wife, Ethel. Data for both families were examined. The data were documented in the form of episodic descriptions of important and/or embarrassing incidents in the families' lives. Each episodic description was summarized by a descriptive title. We computed the number of times each individual was mentioned (cited) in these titles.

Data (Results)

Note: The data are organized by academic rather than by calendar year.

1951–52: 35 episodes; 16 citations
Lucy:	9(56% of citations)
Ricky:	4(25% of citations)
Fred:	1(6.3% of citations)
Ethel:	2(12.5% of citations)

1952–53: 31 episodes; 17 citations
Lucy:	12(70.6% of citations)
Ricky:	4(23.5% of citations)
Fred:	1(5.9% of citations)
Ethel:	0

1953–54: 31 episodes; 17 citations
Lucy:	6(35.3% of citations)
Ricky:	6(35.3% of citations)
Fred:	2(11.8% of citations)
Ethel:	3(17.6% of citations)

1954–55: 30 episodes; 11 citations
Lucy:	4(36.4% of citations)
Ricky:	4(36.4% of citations)
Fred:	1(9.1% of citations)
Ethel:	2(18.2% of citations)

1955–56: 26 episodes; 17 citations
Lucy:	14(82.4% of citations)
Ricky:	3(17.5% of citations)
Fred:	0
Ethel:	0

1956–57: 26 episodes; 24 citations
Lucy:	14(58.3% of citations)
Ricky:	2(8.3% of citations)
Fred:	1(4.2% of citations)
Ethel:	1(4.2% of citations)
Little Ricky:	6(25% of citations)

Interpretation

Historians have tended, wrongly, to see the period 1951 through 1957 as a sociologically dormant era. It was in fact a time of turbulent, dynamic upheavel in the United

The subjects of the study. Left to right: Lucy, Ethel, Ricky, Fred. (Not shown in picture: little Ricky).

LETTERS TO THE EDITOR

He Loved Lucy

Sir:

I am writing with regard to Wilson Kosmowski's questionable research findings on "Titular Dominance in *I Love Lucy.*" I was a neighbor of the Ricardos and the Mertzes. From 1953 through 1955 I had a torrid love affair with Lucy Ricardo. She was more woman than Ricky deserved, but that is neither here nor there. I believe that the fact of this extramarital (and may I add very passionate) romance, of which I am extremely proud, may help shed light on Kosmowski's questionable claim that Lucy played a "less dominant role" in her family's life during those years. In 1953–55 was Lucy less visible to peeping-Tom neighbors? Very likely. But was she less dominant? Hardly. I believe I know which of our neighbors collected the data upon which Kosmowski bases his conclusions. Believe me, that person was not an unbiased observer.

Reginald Penna
Flatbush, New York

He Loved Lucy, Too

To the Editor:

I am greatly disappointed that you chose to publish Reginald Penna's letter (*JIR* 1991; 36:2). Mr. Penna was indeed a neighbor of ours during the 1950s. My personal memories of Mr. Penna are necessarily vague, as I was a very young child at the time (see "Titular Dominance in '*I Love Lucy,*'" *JIR* 1991; 36:1). As to the nature of Mr. Penna's relationship with my mother, that is a matter unsuitable for public discussion in a scientific research journal. Readers who wish to examine the relevant sociologic facts would be better served by consulting my recent book *Lucy and the 1950s Shift in World Culture.*

Richard Reginald Ricardo, Jr., Ph.D.
New School for Social Research
New York, New York

Insufficient Irreproducibility

To the Editor:

My research on titular dominance in *I Love Lucy* (*JIR* 36:1) has inspired numerous other investigators to attempt to replicate my work. They have demonstrated—to my chagrin—that my results were not irreproducible. I therefore wish to withdraw my paper. Please accept my apology. This has been a most humbling experience.

Wilson Kosmowski
Hong Kong

P.S. Interested readers may wish to consult some of the confirmatory reports that have appeared in other scholarly journals in recent months. Among the most notable are Prof. Gayle Gordon's First-hand titular analysis of Lucy (*Scientif Amer*, May 1991); Dr. William Frawley's Deconstructing Babalu (*Ling/Sociol J*, July/August 1991); and especially Prof. Vivian Vance's Perceptions of Lucy: Cutting Off Our Fates to Spite Arnez" (*Chicago Phil Rev*, April 1991).

States. As the data make clear, the period began with women playing a dominant role in community life. During the middle years, men assumed near-parity with women. By the end of the era, though, women had regained—and indeed surpassed—their original level of dominance. A close examination of the data pertaining to Ricky and little Ricky further indicates that by 1957, adult American males had lost prominence to such a degree that their positions in society were being usurped by young children.

REFERENCES

Andrews, B. *Lucy & Ricky & Fred & Ethel: The Story of "I Love Lucy."* New York: Dutton, 1976.

1. A citation of "the Ricardos" was counted as one citation each for Lucy Ricardo and Ricky Ricardo. After little Ricky Ricardo was born, a citation of "the Ricardos" was counted as one citation each for Lucy, Ricky, and little Ricky.

2. A citation of "the Mertzes" was counted as one citation each for Fred Mertz and Ethel Mertz.

3. A citation of "little Ricky" was *not* counted as a citation for Ricky.

4. A citation of "Ricky" was *not* counted as a citation for little Ricky.

5. During the final year of the study, neither Fred nor Ethel Mertz was cited individually. They were cited (and counted here) once each only because the 1956–57 data contained a citation of "the Mertzes" ("Lucy Misses the Mertzes," February 11, 1957). Because of the paucity of data pertaining to the Mertz family, the results concerning the Mertz data are statistically less significant than the results concerning the Ricardo data. The overall trend, however, is clear.

6. During the final year of the study, little Ricky Ricardo rose to prominence. His father Ricky was never cited individually that year. Ricky was cited (and counted here) twice only because the 1956–57 data contained two citations of "the Ricardos" ("The Ricardos Visit Cuba," December 3, 1956; "The Ricardos Dedicate a Statue," May 6, 1957). It is possible that the family's anonymous (and likely unsuspected) chroniclers simply lost interest in Ricky as his sociologic importance waned.

IN MEMORIAM: DEATH OF THE ONE-MINUTE MANAGER

Steven Drew

Microwave Accident in His Home

Dr. Donald G.M. Raker, whose lengthy career consisted of numerous brief but spectacularly productive stints as chief administrator of many of the best known research facilities in the United States, is believed to have died when a household appliance exploded as he was preparing lunch with his wire-haired terrier, Danny Boy.

According to former colleagues, Dr. Raker had taken in recent years to practicing increasingly exotic—some say bizarre—diets and health regimens. Danny Boy was thought to have been an integral part of Dr. Raker's newest diet.

Dr. Raker had been a recluse since his release from Federal prison in 1981, where he served a 1-year sentence for financial activities connected with President Nixon's 1972 reelection campaign.

From 1947 through the late 1960s, Dr. Raker oversaw a long string of research teams that produced many celebrated scientific and commercial successes, among them the infrared strobe light, the high-speed palladium arsenide semiconductor switch, and Jell-O™ brand instant pudding.

It was Dr. Raker's wish that his remains be cremated using a special device of his own design. He will be mist.

LETTERS TO THE EDITOR

Wisdom and Advice for the New Editor

Dear Marc:

I've been with *JIR* since about 1961. In my judgment, *JIR* has become too gentrified, too domesticated, too civilized. The former editor is obviously an English major or some other sort of pervert. When you think of a *JIR* reader, think of a bright graduate student: interested in science, work, grants, and sex, in about the reverse order.

I had two really dirty cracks in my last article, and they took them out. Leave them in. These are scientists you are dealing with, not writers. I once went down to the computer room of the University of Cincinnati Medical School, back in 1968 or so, where they had a huge IBM mainframe. The walls were lined with machines in which huge rolls of tape were rotating. One of them was labeled, in Hebrew, "KOSHER," another, right next to it, also in Hebrew, "TREF." *That* is how real scientists keep their data straight. Get *JIR* out of the literary business and back into the mad-scientist world.

Stanley A. Rudin
Lima, Ohio

Rejection Is Hard to Take

To the Editor:

I broke the Olympic high-jumping record when I read your April 1990 form letter (copy enclosed), in which you accepted my paper, Statistics: Toward a Kinder, Gentler Subject, for publication. When the November/December 1990 issue containing my article arrived, it was the high point of my life. Before the day was over, I had sent autographed copies to all 200,000 statisticians in the world.

Given that astronomic high, you can imagine my nuclear-crater low when I received your recent form letter (copy enclosed) rejecting my article. At first I thought that you must be referring to another of my many papers. But then I realized that this was the only paper I had written since that D-effort in third grade, What I Did Over My Summer Vacation. So now I am puzzled. What will happen now? Will you recall that issue and "white-out" my article? And how would I get those 200,000 copies back? Must I tell my tenure committee?

Sherman Chottiner
Syracuse, New York

NONCONFORMISTS: THE 10% RULE

John Follman
Tampa, Florida

Introduction

In U.S. Marine boot camp in San Diego in October 1950, my lead drill instructor (DI) had a catch-all category of Marine recruits whom he described as "sick, lame, and lazy." This group included those who "didn't get the word," "goofed off," "doped off," were "sick," "didn't get with the program," "didn't conform," "malingered," and so on. The DI, from his considerable experience, had concluded that this group constituted about 10% of his complement of recruits in particular and perhaps people in general.

Since that time this author, wondering about the veridicality of this uneducated man's perception, has been collecting evidence, often anecdotal, that the incidence of people who don't conform is about 10%. Below is a compilation of this evidence, such as it is.

Incidentally, of all humans who have ever lived, 9% live now (Westing, 1981).

Cognitive

IQ
• In the Guidance Study of the Berkeley Survey Group, Honzik, Macfarlane, and Allen (1948) found that the IQs of 9% of the sample changed 30 or more points over time.

• The saliency of the 10% rule is also evident in that 10% was the apparent criterion of failure on the *Armed Forces Qualification Test.* Curiously, those testees scoring below 10% totalled 11.5% (Rohwer, 1970).

EDUCATION
• On any given day, 10% of the approximate 130,000 Hillsborough County (Tampa) public school students are absent (Weyen, 1988). Larson (1956) reported that at that time, 10% of pupils had experienced nonpromotion before the completion of their elementary education.

• According to a press release associated with a book by the American Association of School Administrators, "approximately 10% of today's public school teachers are incompetent" (Elam, 1979). Empirical confirmation of this 10% estimate also emanates from Haney, Madaus, and Kreitzer (1987), and from Sirota (Copelin, February 16, 1984).

Affective

VALUES
• Of Britons polled by *Options* magazine, 10% considered their pets more important to their happiness than their spouses (*St. Petersburg Times*, January 13, 1987).

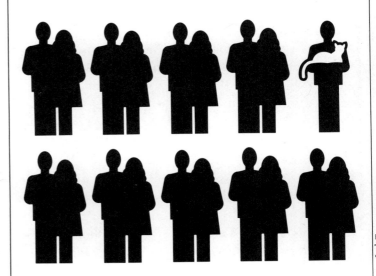

ATTITUDES
• In a poll, about 7% of Austrians were self-declared anti-Semites, while about a third of all Austrians were generally prejudiced against foreigners (*St. Petersburg Times*, March 17, 1987).

Jack Tom

TASTES

• Only 10% of children did not rate ice cream first in a food favorites poll (**St. Petersburg Times**, March 15, 1984).

OBESITY

• According to Bray (1976), at least 10% of American children suffer from obesity.

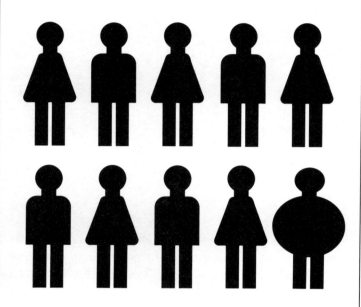

MENTAL HEALTH

• The rule of thumb of the incidence of people suffering from traditional psychological problems is between 10% and 15% (Wright, 1979). Apparently, until recently, seatbelt usage was in the 11% to 12% range (Vick, January 4, 1987).

CHILDREN

• Psychiatric disorders in children usually account for less than 10% of children treated at psychiatric institutions (Werry, 1972); another estimate is 15% (Snapper and Ohms, 1977). Of children, 10% are estimated to be bed-wetters, 10% are hyperatctive, and 15% suffer from bruxism (Wright, 1979). The rate of mental impairment of New York City African-American and Hispanic children was 16% (Shearer, 1969).

• According to Kagan (Hostetler, 1988), 10% of 2- to 3-year-old children who experienced normal births showed shyness, restraint, and subdued affect when presented with unfamiliar people and situations.

• In their early diagnostic study, 10% of children were classified as "difficult" by Thomas, Chess, and Birch (1970). Finally, psychologists estimated that as many as 10% of American children are friendless (Talan, 1987).

IN BETWEEN: TEENS

• In 1981, high-school-age students were responsible for 11.3% of all criminal activity (Leigh and Peterson, 1986).

• It is estimated that only 10% to 15% of teenage drivers are responsible for the bad reputation of teenage drivers (Martin, 1987).

• At least 15% of teenagers aged 16 to 19 are already disconnected from society and unlikely to become productive adults (Vobejda, 1985).

ADULTS

• It was estimated that up to 10% of women in college suffer from bulimia (Foster, 1983); 10% report going on eating binges as often as twice weekly (Litke, 1987); and about 10% of University of South Florida women have an eating disorder (anorexia, bulimia, or bulimerexia) (Carlton, 1986).

• The percentage of Seattle area residents admitting to having attempted suicide was 10% to 12%. (**Tampa Tribune**, February 19, 1984).

• It is estimated that between 10% and 12% of Americans are dental phobic and *never* go to a dentist (Sobel, 1979).

• According to management consultant Robert Bramson, "only" 10% of office workers are relentlessly difficult (**Time**, March 17, 1980).

• On the measure of fear and apprehensiveness of social uncertainty, 10% of men (and 1% of women) chose the frightening and embarrassing alternative over a merely onerous alternative in more than three-fourths of the items (Lykken, 1982).

• Men vs women reported themselves as unhappy in three different situations, that is, at work, 7% vs 12%; at home, 8% vs 9%; and at the beach, 7% vs 6%, respectively (Cameron et al., 1970).

Jack Tom

"NEW" MEN

• Telephone interviews of 1,870 people indicated that 91% of married women reported that they did the shopping, and also that 90% of married women who work full-time reported that they did the primary shopping (Burros, 1988). Beck (1986) reported that 5% to 10% of men are "new" men, that is, men who are willing to support women's request for independence and equality with more than lip service.

DRUGS

• An aggregate of about 22% of the U.S. Army personnel guarding nuclear and chemical weapons sites are convicted felons, and/or alcohol abusers, and/or drug abusers, and/or mentally or medically disabled (Roland, 1986).

• Testing of 300 volunteer truck drivers turned up marijuana in the urine of 14% (**Tampa Tribune**, March 10, 1987).

Jack Tom

• Of 330 National Football League players attending the league's combined draft workout in the week of January 25, 1987, in Indianapolis, 29 (9%) tested positively for drugs (the **Sporting News**, March 16, 1987).

Other

RELIGION

• Of Oral Robert's television ministry 10%—*only* 10%—reported in a poll that they believed him when he said that God had told him to raise $18,000,000 by March 31, 1987, or he would be "called home" (**St. Petersburg Times**, March 31, 1987).

• According to Cleveland Amory (1988), 5% of Americans believe that the afterlife will be boring.

PHYSICIAN, HEAL THYSELF

• An analysis of doctors listed under medical specialties in the 1983 Hartford, Connecticut, telephone directory yellow pages indicated that 12% were not board-certified (**St. Petersburg Times**, May 21, 1987). In another study, 5% of doctors who applied for work at a nationwide chain of outpatient clinics lied about their credentials (**St. Petersburg Times**, February 11, 1988).

• In a random survey of 500 physicians in a state in New England, 10% reported using a psychoactive drug recreationally at least once a month (McAuliffe et al, 1986). Additionally, it was recently reported in a medical journal that alcoholism affects 8% to 10% of American doctors at some time during their careers (Dunn, 1988). One psychiatrist, Dr. Thomas E. Bittker, asserted that 1 of every 10 doctors needs therapy (Bittker, 1984). Dr. Bittker also asserted that 300 doctors commit suicide each year.

• Of male psychiatrists who responded to a survey, 6.4% admitted having sexual relations with their patients (Gartrell, 1987). In another poll, fewer than 10% of responding psychiatrists indicated reporting their colleagues for unethical sexual behavior (Gartrell, 1987).

SPORTS (TIP OF THE ICEBERG)

• Dick Schultz, director of the National Collegiate Athletic Association (NCAA), stated that about 12% of the more than 1,000 NCAA member schools are under investigation annually (Zucco, 1988). Coincidentally, the NCAA has 12 investigators.

CRIME FIGHTERS

• In Chicago, 10% of police officers committed a felony in the presence of an observer. In Washington, D.C., 16% of policemen assigned to high crime-rate areas drank alcohol on the job; in Boston, 8% drank on duty (Laine, 1971). In 1986, Miami Police Chief Clarence

Dickson predicted that about 10% of his force of approxiately 1,000 officers was corrupt. Since then, some 94 officers have been either accused of corruption or convicted (*The Miami Herald*, February 27, 1988).

DOUBLE AGENTS

• About 10% of World War II German agents were double agents working also for the United States (Ahern, 1982).

PUBLICATIONS

• About 10% to 15% of the scholars and scientists produce about 50% or more of all scholarly and scientific publications (Follman, 1984). Conversely, it has been estimated that only 10% of scholarly publications are really scholarly (Priestly and Kerpneck, 1977).

Conclusion

If the number 7 is a lucky number, perhaps the number 10 is an unlucky number. In Asia, the characterization of anything as "number 10" is definitely undesirable.

Incidentally, the actual incidence comprising the catch-all category of Marine recruits who apparently are "sick, lame, and lazy" is 12%. Novaco, Cook, and Sarason (1983) reported that 88% of all recruits who begin Marine boot camp complete it, while the remaining 12% are discharged for a variety of medical, psychologic/behavioral, and other reasons.

REFERENCES

Ahern T. Document: 10% of German agents helped allies. *St. Petersburg Times* Dec 22 1982; p. 9A.

Amory, C. The best and worst of everything. *Parade* Jan 3 1988: p. 10.

Beck J. New woman: Men prefer older model. *Tampa Tribune* Sept 4 1986: p. 19A.

Bittker T. Specialist: 10% of doctors need therapy. 1984, Dearborn, Michigan: Michigan Medical Society.

Bray G. The obese patient, In: *Major Problems in Internal Medicine* (vol. 9). Philadelphia, PA: W.B. Saunders, 1976.

Burros M. Are couples beginning to share household chores equally? HA! *St. Petersburg Times* Mar 1 1988: p. 1D.

Cameron P, Biber H, Brown N, Siro M, Galden C. Consciousness: Thoughts about world and social problems, death, and sex. Kentucky Psychological Association, Sept 25 1970.

Carlton S. (1986, October 1). On campus eating disorders hit close to home. University of South Florida *Oracle* Oct 1 1986, 22, Insight, 2–3.

Copelin L. Educators mark satisfactory grade for own schools. *Austin American Statesman* Feb 16 1984: p. 1.

Doctors' yellow pages listings called misleading, *St. Petersburg Times* May 21 1987; p. 4A.

Dunn MD. Physicians take stand on shaky pedestals. *Tampa Tribune* May 19 1988: Section I, 1, 3.

Elam SM. Some observations on incompetence. *Phi Delta Kappan* 1979; 60:337.

Follman J. Publish or perish: Myth or matter. Fourth Annual Conference on Teaching and Learning in Higher Education, June 19 1984, London, Ontario, Canada.

Foster S. "Binges" common among coeds, doctor says. University of South Florida *Oracle* 1983: p. 1.

Gartrell Reporting practices of psychiatrists who knew of sexual misconduct by colleagues. *Am J Orthopsych* 1987; 57:287–295.

Haney W, Madaus G, Kreitzer A. Charms talismanic: Testing teachers for the improvement of American education. In: Rothkopf E (ed) *Rev Res Educ* 1987; 14.

Honzik MP, Macfarlane JW, Allen L. The stability of mental test performance between two and eighteen years. *J Exper Educ* 1948; 17: 309–324.

Hostetler AJ. Children biased to be bold, shy. *Monitor* April 1988: p. 6.

Ice Cream Ranks No. 1 with Kids. *St. Petersburg Times* Mar 15 1984; p. D1.

Kilpatrick JJ. Old values need to be restored. *St. Petersburg Times* Dec 3 1986: p. 15A.

Laine P. Researcher claims number of police commit crimes. *St. Petersburg Times* May 26 1971: p. 16A.

Larson RE. Age-grade status of Iowa elementary school pupils. In: Coffield WH, Blommers PC. Effects of nonpromotion and educational achievement in the elementary school. *Jour Educ Psychol* 1956; 47:235–250.

Leigh GK, Peterson GW. (eds) *Adolescents in Families*. Cincinnati, Ohio: South-Western, 1986.

Litke J. Survey disputes bulimia "epidemic"? *Tampa Tribune-Times* Sept. 6, 1987: p. 10A.

Lykken DT. Fearlessness: its carefree charm and deadly risks. *Psychol Today* 1982; 16:20–28.

Martin ST. Distractions and drinking among the dangers for teens. *St. Petersburg Times* Dec 20 1987: 1A, 12A.

McAuliffe WE, Rohman M, Santangelo BA, et al. Psychiatric drug use among practicing physicians and medical students. *N Engl J Med* 1986; 315:805–810. Police on Trial. *The Miami Herald* Feb 27 1988: p. 22A.

Novaco RW, Cook TM, Sarason IG. Military recruit training. In: *Stress Reduction and Prevention*. New York: Plenum Press, 1983; 377–418.

Poll: Many Britons prefer pets to spouses, children, money. *St. Petersburg Times* Jan 13 1987; p. 8A.

Poll: Austria is 7% Anti-Semitic. *St. Petersburg Times* Mar 17 1987; p. 7A.

Poll: TV ministers viewed unfavorably. *St. Petersburg Times* Mar 31 1987; p. 4A.

Priestly FEL, Kerpneck H. Publication and academic reward. *Scholarly Pub* 1977; 8:233–237.

Rohwer WD Jr. Cognitive development and education. In: Mussen PH (ed). *Carmichael's Manual of Child Psychology*. New York: John Wiley & Sons, 1970.

Roland N. Army audit finds criminals guarding nuclear, chemical weapons. *St. Petersburg Times* Aug 10 1986; 5A.

Rosenberg M. Inventing the homosexual. *Commentary* 1987; 84:36–40.

Scott W. Personality Parade. *Parade* Sept 21 1986; p. 2.

Shearer L. Poor kids. Intelligence Report. *Parade* June 1 1969, p. 4.

Shearer L. Intelligence Report. *Parade* Oct 26 1986; p. 17.

Snapper and Ohms. (1977).

Sobel D. Researchers try to reduce fear of the dentist. *St. Petersburg Times* Dec 19 1979; p. 15A.

Under a blanket of suspicion. *The Sporting News* Mar 16 1987; p. 40.

Study: Doctors lied on resumes. *St. Petersburg Times* Feb 11 1988; p. 7A.

Study: Suicide "the rule" in the U.S. *Tampa Tribune* Feb 19 1984; p. 22A.

Talan J. Why are some kids friendless? *Tampa Tribune* Mar 30 1987; p. 1AA.

Thomas A, Chess S, Birch HG. The origin of personality. *Sci Amer* 1970; 223:102–109.

Troublemakers in the office. Behavior. *Time* Mar 17 1980; p. 72.

Vick K. Club members don't need buckle-up reminders. *St. Petersburg Times* Jan 4 1987; p. 1B.

Vobejda B. Teens at risk. *St. Petersburg Times* Nov 3 1985; p. 16A.

Werry S. The childhood psychoses. In: Quay HC, Wemy JS (eds) *Psychopathology Disorders of Childhood.* New York: Wiley, 1972.

Westing AH. Note on how many humans that have ever lived. *BioScience* 1981; 31:523–524.

Weyen W. Tracking truants a vexing task. *St. Petersburg Times* May 9 1988; p. 1.

Wright L. Health care psychology. *Amer Psychol* 1979; 34:1001–1006.

Zucco T. NCAA's Schultz waging one-man crusade. *St. Petersburg Times* Ap 7 1988; pp. 1C, 10C.

NOBEL THOUGHTS

Profound insights of the laureates

Marc Abrahams

Eric Chivian is a staff psychiatrist at the Massachusetts Institute of Technology and Assistant Clinical Professor of Psychiatry at Harvard Medical School. He is one of four American and three Russian cofounders of International Physicians for the Prevention of Nuclear War. The organization was awarded the Nobel Peace Prize in 1985. We spoke in Dr. Chivian's office in Cambridge, Massachusetts.

Is your desk neat or messy?

Very messy. Chaotic. I have pictures of my lady friend, my boat, and my kids; papers to get to, bills to pay. I periodically—about once a week—wade through it, straighten it up, and make resolutions to keep it more organized.

What characteristics are most important in choosing a comfortable chair?

For me, I use a rocking chair, and my patient also has a rocking chair. That's nice because it's very hard to sit still for 50 minutes, either for me or for the patient. It's kind of nice to be able to move around. I think rocking chairs are great inventions, anyway. I also use a pillow for my back. Hours and hours of sitting are an occupational hazard for psychiatrists.

How do you take your coffee?

I have a slight coffee fetish. I buy my coffee from The Coffee Connection. I mix decaf with regular. I buy fairly strong roasted coffees, and mix them half and half. I'm very particular about making my coffee. I heat my water, which is spring water—this sounds very peculiar—in a Pyrex florence flask. I use a Melior, which is a glass coffee maker that suspends the coffee and has a screen plunger to push down the grains. I add sugar and use condensed milk. I would use cream but I'm trying to keep my cholesterol down and condensed milk doesn't cool it too much.

Do you have any advice for young people who are entering the field?

One thing is that you don't have to know everything. When you go to medical school there's a sense that you have to be aware of everything or else somebody's going to die. I don't think that's true. Unless you're going out in the Sudan, you're surrounded by lots of other people. I think it's important to really learn well the stuff you're most interested in, and learn the rest as well as you can. But you've got to maintain some perspective or else you'll go crazy. Because there's just too much to know, especially now.

The other thing is that medicine, especially psychiatry, can be an extremely lonely profession. There's the sense of intimacy, because you're with people in a very intimate way all day, but it's only in one direction. And not only that, but you can't really talk about what you do to very many people, even to your significant other or to your family. You can't go home and say, "Guess who I just saw." That's hard. So you really need a lot of collegial interchange.

IN MEMORIAM: ACCUSATIONS HAUNT THE LEGACY OF DR. BRUTTELHEIM

Stephen Drew

When Dr. Benno Bruttelheim took his own life last year at the age of 86, the obituaries stressed the greatness of the man and his pioneering methods. But the praise has now led to a torrent of criticism, with the psychiatrist portrayed not as a dedicated man of wisdom but as a megalomaniacal tyrant who systematically abused patients and undermined their self-confidence.

The opening salvo came in a book (*The Mis-Uses of Enchanting Color*) by Michael Johns. Johns, who was a patient of Bruttelheim from 1966 to 1973, charges that Dr. Bruttelheim's acclaimed practice of using environmental color to influence patient's moods was often taken to cruel extremes.

Johns says that Dr. Bruttelheim would suddenly become enraged at patients, and cover them with latex paint. "He bullied, awed, and terrorized the patients at his clinic, their families, the staff members, his graduate students, and anyone else who came into contact with him," writes Johns.

Another former patient, Donald Board, said in an interview that he was covered with paint some 20 times during an 8-year stay at the clinic. "He rollered me with blue paint when I was suffering depression, brown paint when he felt I was being lethargic, and black paint when I was suicidal," Board said, adding that Dr. Bruttelheim "covered his tracks by claiming to be color-blind."

The attacks have not gone unchallenged. Defenders of Dr. Bruttelheim have come forward, many of them former counselors at the clinic which he directed from 1944 to 1973. "I never saw any of that kind of behavior some of these former patients are reporting," said Karin McNeil, a psychotherapist at the clinic from 1956 to 1964. "Dr. Bruttelheim only used soft pastel colors. In my experience, this was not inconsistent with the spirit of loving and concern for the patients' mental health and appearance."

Ilko O'Brien-Hnkmunnim, a former lay counselor at the clinic who was married to Dr. Bruttelheim for nearly a month during the 1960s, said that when Bruttelheim rollered patients he was always acutely aware of the purpose he had in mind. "Benno was the Jackson Pollack of the human psyche," she said. "Abnormal behavior was his canvas."

Dr. Louis Festo, who was chief administrative surgeon at the clinic until 1969, said that "Dr. Bruttelheim abhorred touching human skin. He was nauseated by it. I don't want to say that these former patients are making these things up, but these stories appear to be the normal psychologic adaptations of severely disturbed people who were rollered with paint 10, 20, or 30 years ago."

What is at stake in all of this conflicting testimony is the standing and credibility of one of the century's most important scholarly and therapeutic legacies. Bruttelheim, after all, can no longer speak for himself, so there is no way to resolve the conflicting stories of a man everyone sees as not only powerful and complex, but colorful as well.

CHAPTER 8

HISTORY AND ITS FUTURE

PATTERNS OF LIMB RETENTION IN HELLENIC STATUARY

Pandareus von Grundenstein

New Haven, Connecticut

Many people when traveling in Greece or visiting museums cannot help but notice a conspicuous absence of upper limbs in classical Greek statues (Figure 1). As a scientist and a scholar, I felt that there was probably some scientific reason for this dearth of arms, and it became my purpose to discover it. Over a period of 17 years (and several fortunes), I undertook to clear up this sorely neglected branch (or should I say: arm) of science. Armology is truly one of the least-studied yet most fascinating of scientific disciplines. My studies were concerned with only a fraction of the arms in the world, yet they revealed a wealth of new information.

My research was concerned with the arms, or lack thereof, of the most renowned and celebrated classical statues, as well as many less famous works of art. Extensive research and a few simple calculations yielded the following data: In all, 60% of all Greek statues are missing some portion of their upper limbs. To study the changes in limb retention over the centuries, I classified the statues into three groups: Early, Middle, and Late. My data show that 64% of Early statues, 61% of Middle statues, and 55% of Late statues are deficient in arms. Supporting this trend is the fact that Roman statues have a full quota of upper limbs in over 50% of cases (Figure 2). It thus remained for me to discover the cause of this paucity of arms.

Figure 1

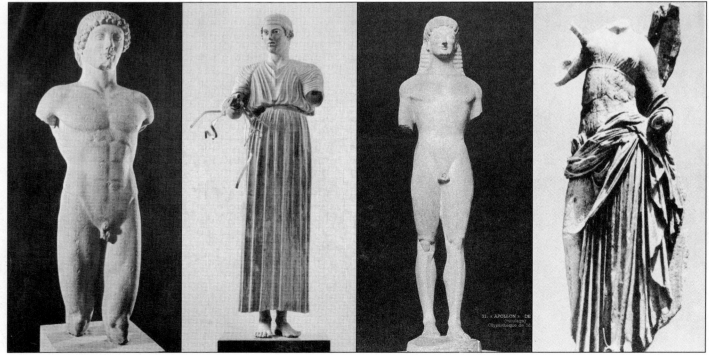

Photos courtesy of the New York Public Library Picture Collection

Several theories have been advanced to date, but none of them explain the facts as adequately as the one that I submit here. The first of these erroneous theories is the "Your Mother Was Right When She Told You Not to Chew Your Fingernails, Because Look What Happened to the Greek Statues" theory, which was propounded by Professor Martin Paxston. The second major theory, advanced by Thomas Grundibus, B.S., M.A., Ph.D., L.L. Bean, is the "Alpha Mu Epsilon" theory. This theory concerns the Disarmeans, a somewhat handicapped tribe of Goths who lacked the upper

Figure 2. Patterns of Limb Retention.

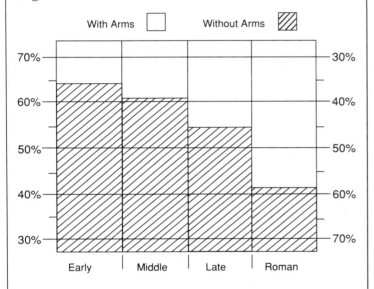

appendages that are generally considered standard issue on *Homo sapiens*. Dr. Grundibus maintains that the Disarmeans attacked the Greeks and, out of jealousy, kicked the arms off all the sculptures. Finally, the last of these popular but erroneous theories, suggested by Dr. Kennard Willbram, makes the assumption that the arms fell off merely because the statues were old and were knocked about a good bit. There are obvious drawbacks to each of these theories.

For Professor Paxston's theory, fingernail-biting is clearly not an activity of cultured people, yet the Romans, who were demonstrably far less cultured than the Greeks, have more arms per statue. The first theory is, therefore, not supported by the facts. The second theory was disproved by Herr Professor Jolk Danielson, who, after tedious and careful archeologic studies, has conclusively shown that those somewhat disabled bar-

barians the Disarmeans never penetrated as far south as the Hellenic regions. Finally, Dr. Willbram's theory is obviously inaccurate, because if rough treatment were the cause of the deficiency of arms in sculptures, it is apparent that the legs would be absent as well.

At this point the careful reader may well be wondering about the theory that I have developed to explain the scarcity of upper appendages. It is patently manifest that there are a finite number of arms in the world, and therefore if arms are missing from one place, they must be present at another location. The Law of Conservation of Matter ensures as much. The difficulty is in ascertaining where the missing limbs may be found. There are admittedly a vast number of locations in the world, but I set myself the task of discovering in which of these locations a noticeably sizable accumulation of arms is apparent.

After long travail I reached my goal: India. Upon discovering the multitude of arms on Indian art (Hindu art in particular), I embarked on an in-depth survey of Indian arms (Figure 3). I discovered that there are, on the average, about three arms to every Indian statue body. I then referred to my data on Greek statues. While 60% of the statues are missing some portion of their upper limbs, some retain half of one arm, some are missing only one hand and wrist, still others have no arms at all, and so on. Using the data regarding these fractions, I calculated that the Greek statues have an average of one arm per body. Putting these facts together and including a slight allowance for unavoidable experimental error, I found that the ratio of upper limbs to torsos is precisely 2:1, which is the predicted norm (Figure 4). Clearly, then, the expected number of anterior appendages indeed exists; the only change is in the location of some 60% of these appendages. Lending strength to the discovery is the observation that few indeed are the arms that are actually missing from Indian sculptures.

Having demonstrated that the misplaced Greek arms may be found on the torsos of Indian statues, we must now seek to explain the method of transport. There are several theories regarding this traffic in arms. It seems reasonable to assume that the arms were captured from the Greeks by force. If we take this into consideration, there is one theory that merits more serious consideration that the other. Here we enter the bailiwick of Professor Rana Formodo, who has made arms distribution her field of study. According to her theory, because of the limited number of arms available to sculptors, all the statue-making nations came together in Helvetia for the SALT (Statue Arms Limitation Treaty) talks. At the talks it was established that each statue-making nation would be limited to a strict quota of lateral limbs. Soon after, however, India felt that it was her natural right to receive a larger quota. Nevertheless, the SALT could not be modified. India executed a series of lightning-quick raids, the purpose of which was to purloin Greek arms. This disarmament constituted such an outrageous treaty violation that it prompted the SALT II talks. Unfortunately, the two arms superpowers were unable to reach an agreement, and the arms race continued until the fall of the Greek Empire, although it abated somewhat toward the end, owing to the watchfulness of the Greeks. Obviously, the Roman military was too strong to allow the Indians to steal as many Roman arms. This, then, is the explanation of the mysterious disfigurement of classical sculpture.

Figure 4. Limbs and Arms: Expectations and Observations.

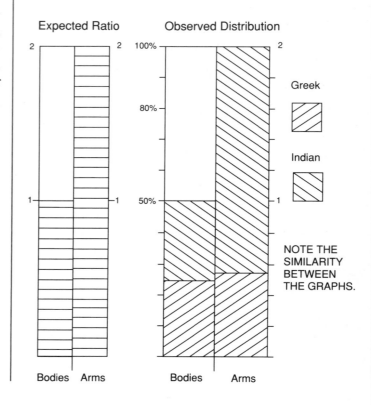

Expected Ratio   Observed Distribution

Greek

Indian

NOTE THE SIMILARITY BETWEEN THE GRAPHS.

Bodies Arms   Bodies Arms

Figure 3

Photos courtesy of the New York Public Library Picture Collection

REDUCING HISTORY TO NUMBERS

Susan Mary Smith
Montreal, Quebec

For too long now, traditional historians have sneered at the reductionist theory of history.[1] Such scholars perpetually refuse to acknowledge the simplicity and succinctness of reductionist history. This situation, which I purport springs mainly from envy, must be halted, if not slowed down.

The conciseness and clean-cut quality of reductionist history (RH) can be used not only by professional historians, but also by first year students and, in exceptional cases, graduate students (excepting, of course, those studying Canadian history).[2] I should like to share one example of how useful RH can be: I choose the much argued question "What was the cause (please note I said *cause* and not causes) of the Second World War?"

Using the standard equation:

$$Rh - A = \frac{c^2 \, \Omega}{\sqrt{-1}} \times \pi$$

we get (cancelling out all the Greek stuff)

$$Rh - A = \frac{C^2}{\sqrt{-1}}$$

Simplifying further, we remove all the numbers:

$$Rh - A = \frac{c \times \frac{c}{x}}{\sqrt{}}$$

Then, taking out the variable integers, we are left with

$$Rh - A = c \times c$$

It is easy to deduce from this now *much* simplified equation that Hitler was born at *precisely* (and precision is the key here) 4:10:33 P.M. (Eastern Standard Time) on June 3 1902.

The cause of WWII is, using the equation

$$Rh = \frac{c \times c}{A}$$

and assuming that $c \times c$ is Hitler's birthdate (3 × 6, June being the sixth month, × 1902)

<_><_>18 × 1902

(sorry, no calculator).

The answer is

$$ARh = 11$$

Funding for the research undertaken in this project was supplied by "Funds for Excellence."

---

1. See Taylor AJP, *Origins of the First World War;* and Hitler A, *Mein Kampf.*

2. See Smith SM, *New methods of RH* (not anything to do with blood or monkeys), Simon Fraser University Press, 1964, pp. 1–2.

SOLUTION TO LAST ISSUE'S PUZZLER

The given conditions of the problem guaranteed that:

(A) there were 79 coconuts when the cannibals were first sighted;
(B) the thief immediately took 26 of the coconuts; and
(C) the thief's twin brother departed on his journey when his watch read EXACTLY 2:00 PM

When the thief later gave one of the 26 coconuts to Diogenes the Monkey (leaving 52 for Gilligan), it became impossible to get enough material to build the dam.

The Skipper's rhyme provides a way to answer the riddle. Thus, "if there is an honest man in the fort, then there are more coconuts on two of the islands."

There WERE honest men still in the fort (the cannibals, remember, had eaten coconuts instead of their usual afternoon snack), so the Skipper was able to build the dam before the evening banquet began. As a result, when the thief's brother returned, the cannibals could have eaten at most 17 honest men.

The correct answer, then, is "17."

PRESERVING THE GRAND CANYON: FINAL REPORT

Earle Spamer
Academy of Natural Sciences, Philadelphia, Pennsylvania

Introduction
Between 1974 and 1990 the author headed a research program to identify the best means of preserving the Grand Canyon. This report concludes the study and makes recommendations for implementation.

The Problem
The Grand Canyon of Arizona is threatened by human and natural factors. Human-generated atmospheric haze, operations of the Glen Canyon Dam power plant, and natural forces of erosion are destroying the canyon.

Present technology is incapable of moderating the erosion of the Grand Canyon. In December 1966, catastrophic storm-generated debris flows delivered tons of the Grand Canyon to the Colorado River (Cooley *et al*, 1977). The National Park Service was unable to prevent this level of massive destruction, and these processes are still a threat.

The Solution
The impending loss of the Grand Canyon calls for an interim means of protection until such time as the technology is developed to save the canyon. The immediate solution is to fill in the canyon.

Materials and Methods
Dirt is, paradoxically, an unacceptable fill for the Grand Canyon. It is heavy, it is hard to keep clean, and it would be difficult to re-excavate from the canyon. But styrofoam piffles (also called peanuts, nerdles, doodles, etc.) of the kind used to cushion articles during shipping may serve very well.

Five kinds of piffles were tested for durability, vol-umetric characteristics, and cost; the piffles are the standard Figure-8, Tree Bark, Cheez-Puff, Potato Chip (un-ridged), and Dome (Figure 1). The control sample was the standard sterile cotton ball. All piffles resist compression, although the Potato Chip variety tended to break in half when compacted. Owing to very low mass, very large volumes of piffles do not compact to a significant degree.

Figure 1. Piffles. *Top row:* Figure-8, Tree Bark, Cheez-Puff. *Bottom row:* Potato Chip, Dome, Cotton Ball (control).

The Grand Canyon is an empty volume of about 4.17×10^{12} m³ (Lucchitta, 1988). This volume is henceforth referred to as "1.0 GC" Each kind of piffle was used to fill a standard 16-fl. oz. jar (473 cm²) and counted. Extrapolating the 16-oz. volume of piffles to 1.0 GC is within statistical limits of error, due to low compactivity. The control sample of cotton balls was measured more directly from a box of 130 balls. As the manufacturer has already determined the optimal packing density of

130 balls, the volume of the box (3.9×10^{-16} GC) was extrapolated to 1.0 GC.

Units of measurement are converted to short tons for practical reasons. Over-the-road haulers will not tolerate exponential figures or measurements in grams, and government shipping regulations are written in English units. Production and shipping costs, although unknown, are estimated here (Table 1). Piffles are no longer manufactured; they are saved and reused. A finite number of piffles was manufactured between 1963 and 1966; the entire supply is currently in a complex cycle of storage and use (Sornovich, in press).

The new production cost is estimated at $0.0001/piffle. Shipping and handling costs are added at an estimated $5,000/truck. Only the control sample was calculated differently. (The control box of cotton balls cost $2.17 at a discount drug store.) Extrapolating, the total cotton ball infill cost is 5.5×10^{15}/GC. The manufacture and shipping cost of piffles is three orders of magnitude less per GC.

Environmental Considerations

Piffles are inert and last forever. Air circulation between piffles is very good, reducing the probability of spontaneous combustion and damage to the canyon by fire. Respiration by indigenous fauna and flora will be unimpeded.

The total load for each piffle type and the control sample was calculated for a unit column of 1.0 square foot by 6,000 feet (the greatest depth of the canyon). By intuition, deformation of the earth's crust is de-

monstrably negligible in all cases; thus no seismic hazard due to isostatic load is anticipated. Loss of surface-layer piffles into the nearby environment will probably be only unsightly and not pose a danger to the ecosystem. Periodic gathering of piffles and returning them to the canyon will efficiently remedy the problem.

Nonbiodegradability of piffles is avoided with the use of cotton balls. However, cotton balls are susceptible to water saturation, reducing porosity through compaction and increasing the probability of decay and spontaneous combustion. The soggy mass would also present clean-up problems. Removal of piffles from the Grand Canyon will be very easy. The Grand Canyon is open to deserts in the west, and the entire load can easily be blown out to Nevada with leaf-blowers.

Empiric Tests

In 1990 the author dropped a randomly selected piffle at the bottom of the Grand Canyon. There was no measurable effect of the isostatic load on the earth's crust, nor was there any observable impact on the Grand Canyon ecosystem. Removal of the piffle from the canyon was accomplished with no difficulty. In a control test, a randomly selected cotton ball was dropped, again with no measurable isostatic impact. It immediately became clogged with sand and small bugs, however, and had to be discarded.

Discussion

Although cotton-ball preservation of the Grand Canyon is more sound in terms of biodegradability, the cost of

Table 1. Materials and costs to save the Grand Canyon.

	Piffle Type					Control Cotton Ball
	Figure-8	**Tree Bark**	**Cheez-Puff**	**Potato Chip**	**Dome**	
No. required	291×10^{15}	229×10^{15}	238×10^{15}	264×10^{15}	326×10^{15}	331×10^{15}
Total weight kg tons	12.4×10^{12} 13.6×10^{9}	17.1×10^{12} 18.9×10^{9}	16.5×10^{12} 18.2×10^{9}	12.3×10^{12} 13.6×10^{9}	7.7×10^{12} 8.4×10^{9}	143×10^{15} 157×10^{9}
No. of trucks	568,515,367	786,490,743	756,685,947	565,360,993	351,631,891	6,556,043,838
Total cost ($)	31.9×10^{12}	26.8×10^{12}	27.6×10^{12}	29.2×10^{12}	34.4×10^{12}	5.5×10^{15}
Weight of 1 ft² column 6,000 ft high (tons)	0.53	1.62	1.34	0.89	0.91	6.41

Figure 2. Artist's rendering of the Grand Canyon at a time near completion of piffle-fill conservation. *Left:* Before. *Right:* After.

piffles is several orders of magnitude less. Tree Bark piffles are the most economical but also the heaviest load to deliver and maintain. Figure-8 piffles are the second most expensive, but lightest, kind of storage medium. Their optimal shape, providing empty space within the unit cell even after moderate compaction, makes them the most effective means for *in piffum* storage of the canyon.

To offset loss of income from tourism, government subsidies are anticipated.

REFERENCES

Cooley ME, *et al.* Effects of the catastrophic flood of December 1966, north rim area, eastern Grand Canyon, Arizona. *US Geol Survey, Prof Paper 980.* 1977.

Lucchitta I. Canyon Maker. *Plateau* 1988; 59(2):

Sornovich AY. Are piffles the missing mass? *J Irreprod Astrophys,* in press.

ABRAHAM LINCOLN'S MUSTACHE

Don Dorrance
South Milwaukee, Wisconsin

As I have indicated throughout this modest effort, my major field of investigation is music. It is a peculiarity of this discipline that academic papers use "notes" that are, indeed, notes. Thus, I have, in another connection, referred to a major theme of the Jupiter, referenced as:

The fact that I admittedly am outside my field creates a degree of uneasiness, in that the "notes" are of a different definition. I can only promise to do my best.

While doing original research for my dissertation, "The Influence of the Industrial Revolution on the History of Music, with Particular Reference to the Application of Steam," I have come across a curious—and perhaps significant—error in a related field. It is my earnest hope that a qualified, in-the-field researcher will come to my aid.

I came across the first indication by accident while going through a ***Harper's Weekly*** dated March 17, 1860.[1] While my major concern was an article on circuses, referenced by Plotkin, I could not help but notice a political article entitled "Leading Candidates." I was, at first, amused by the description of "A. Lincoln" as being "tall, ungainly, with deep brown eyes above a full mustache." I took little note of this description at the time.

It was 2 weeks later, while checking the ***Leslie's Weekly*** article (May 28, 1862) on characteristics of trumpet music on modern [sic] battlefields—based on an episode at, according to the article, Antietam—that I was struck by an engraving[2] of the then President, Abraham Lincoln, at the telegraph. This engraving (Figure 1) indicated that President Lincoln had a walrus mustache. I noticed this, but decided to let it pass. (Music, rather than history, is my field of interest.)

Three months later, while trying to verify a reference to the steam calliope, I noted a group portrait entitled "Lincoln and His Generals" (***Scribner's,*** March 1863)[3] in which, once again, the President (Lincoln) was depicted as mustached.

Figure 1

Photo courtesy of the New York Public Library Picture Collection

At this point my curiosity was aroused enough to consult Sandberg's Abraham Lincoln. According to the Index there were 27 references under the generic term, "Beard." I carefully checked each page referenced, as carefully as I would check each note in music, and Sandberg (Carl) uniformly maintained that Abraham Lincoln had a beard but no mustache.

I was, as you may imagine, suffering from an aroused curiosity. Sufficiently, indeed, to put aside for a moment my original research, which is in music. I began a rapid testing of the evidence, as best I could working in a strange (to me) field, that of history. I determined after some thought that a "quick and dirty" survey might best be conducted through textbooks, as, to the best of my recollection, it was in school that I had formed my first impression of Lincoln's face.

At length I discovered a complete set of McKinney's Eclectic Readers for Grade 6, editions of 1867, 1871, 1873, 1877, 1880, and 1882.[4] In this series of readers, Abraham Lincoln is depicted with a mustache through 1877. In the 1880 edition he is clean-shaven; in 1882, bearded. But not mustached.

I decided at this point I was out of my depth and determined to enlist professional help, being aware that I was out of my field of authority. I requested aid from the American Historical Association of Philadelphia. My request was refused.

A series of letters followed, which, you may well mark, will be reprinted in their entirety at the conclusion of these investigations. I can only at this time (pending future and complete publication) state that the American Historical Association has not shown the curiosity I would expect of a respected science, that of history.

In seeking clarification without their aid, I was able to find an indication. It unfortunately proved to be negative. But this negative has enormous implications. For the minutes of the May, 1879, meeting of the American Historical Association are missing. Think of this what you will, but repeated inquiries on this subject brought the weak response that there was in fact no meeting that month.

A most significant question arises from an examination of the "facsimile" edition of **Harper's Weekly** published from 1960 to 1965 purporting to be an exact duplicate of the famed Civil War-era newspaper. A careful comparison of the facsimile has Lincoln with a beard, and in every case in the original he has a mustache. I find this hard to credit as coincidence.

The facts, to me, have become puzzling. Abraham Lincoln, until May 1879, had a full mustache. After May 1879, Abraham Lincoln was clean-shaven. And Abraham Lincoln, sometime after March 1881, had a beard but no mustache.

Further research has led to tentative conclusions, which I offer to any qualified historian, since my major field of interest is music and I do not feel that I am qualified to assess such potentially important conclusions which are so far from my field, which is music. The secret which I appear to have uncovered, to suggest areas for future investigation, lies in the perplexing problem of the parentage of Chester A. Arthur, 19th President of the United States (Figure 2), especially as referred to in the rigorously ignored article by Stanwell Momper in the **Century Magazine** of February 1882, entitled, "What Became of Anne Rutledge?"[5] This, of course, follows the article by Fenton, "Who was Chester Arthur's Father?"

I wish to return to my true field of interest, which is music. I can only appeal to someone more competent in history, in which I am not an expert, to rise to answer the questions which have been bothering me.

I hope the footnotes are satisfactory. They are not the kind I usually use.

Figure 2

Photo courtesy of the New York Public Library Picture Collection

1. Leading Candidates. *Harper's Weekly* Mar 17 1860.
2. Trumpets on the Battlefield. *Leslie's Weekly* May 28 1862.
3. Lincoln and His Generals. *Scribner's* Mar 1863.
4. *McKinney's Eclectic Readers.* Grade 6. 1867, 1871, 1873, 1877, 1880, 1882.
5. What Became of Anne Rutledge? *Century Magazine* Feb 1882.

Contains 100% gossip from concentrate

Compiled by Stephen Drew

Improving Public Opinion

A new public opinion poll that will be able to gather data more quickly, more often, and much more accurately than current polls will soon be available. The "Typical American Poll" is being organized by several major U.S. public opinion research organizations. It is sponsored jointly by the Republican and Democratic political parties, several professional sports leagues, and a consortium of domestic automobile manufacturers. A panel of 200 statistically typical citizens is being recruited. They will be permanently relocated to the poll's live-in research center in Des Moines, Iowa, where they will be instantly available at all times for polling. The European Economic Community is funding a similar project that will be based in Belgium.

Abundance of Personality

The number of personalities that an accused research grant embezzler says are contained within her psyche increased to eight last month as testimony continued in U.S. Federal Court in Manhattan. In a hostile interchange with U.S. assistant district attorney Edwin Beister, Dr. Ramona Hunt insisted that she has multiple personalities. Hunt argued that her condition was not unusual among people engaged in multidisciplinary research. The trial concerns the disappearance of more than $650,000 in funding that was to be used for studying the role that neutrons play in digesting fatty acids. Hunt was one of five researchers involved in an earlier study concerning the relationship between plate tectonics and gastroenterologic motility.

Waiting for Goddard

A group of aerospace scientists at Louis University in Paris may be nearing the long-sought goal of producing a cheap, low-mass rocket fuel. This development could spark a new golden age of space exploration and ex-

perimentation. The research team, led by Pierre Latour, Lucien Raimbourg, Vladimir Estragon, and Luciano Pozzo, is experimenting with ultra-distilled recycled biomass. The raw material is obtained from urban garbage recycling programs. The key aspect of the biomass conversion process was developed by Dr. Samuel O. Goddard, a lateral descendant of American rocketry pioneer Robert Goddard. The younger Goddard is a biochemist whose work has centered on the analysis of fruit peelings. The project's final stages are awaiting Goddard's return from a 3-year vacation.

SLEEP RESEARCH UPDATE

- SD is sleeping with DB and GW.
- GW is sleeping with LU.
- LU now sleeps with KB.
- KB now sleeps with OG.
- OG continues to sleep with WHF and is maintaining detailed records using electronic media.
- WHF is no longer sleeping at home.
- RR is only sleeping at home.
- PT is sleeping with GF and LE.
- SG has resumed sleeping with members of the Psychology Department.

Voice of Experience

There appears to be a reliable test to identify individuals who are truly experienced at their profession, developed by Marilyn Darling, senior researcher at Signet Network, Ltd. Darling records conversations with the individuals and then searches the printed transcripts for keyword indicators. She has identified a simple indicator that is approximately 87% effective in identifying experienced individuals. According to Darling, "If the person's conversation includes, even once, the phrase 'never again,' then that person is likely to be truly experienced."

Darling originally began the work as part of her research into selecting workers from among a pool of applicants. She says the technique may also be applicable to identifying individuals who are experienced in other domains, such as marriage, parenthood, and draw poker.

Improved Cosmologic Constant

The cosmologic constant will be adjusted on January 1 to bring it into accordance with predicted values. The International Academy of Physical Constants will add .000017 to the current value. The cosmologic constant last underwent adjustment on January 1, 1968.

Extended-Wear Chewing Gum

What chews around comes around. If you chew a piece of gum for more than 5 days, it begins to have a "second wind" of flavor, it has been discovered. Roberto Kurosh-Coltin of the Rund Corporation has been conducting gum studies since 1963. Kurosh-Coltin has not identified the chemical changes underlying the phenomenon but says that it may be somewhat analogous to the stress fatigue exhibited by metals undergoing prolonged intensive use in heavy mechanical equipment. The "second wind" flavor is usually different from, although perceptually related to, the gum's original flavor. In some types of gum, a fermentation process may be involved.

Placebos: Good News

The severe shortage of reliable placebos may soon be a thing of the past, according to a report published by The Placebo Institute. "The bioengineering revolution is giving us new ways to synthesize previously rare materials," explained Jamie Watt, the institute's Director of Research. "I hesitate to use the word, but it seems almost miraculous." However, Watt cautions that some shortages will still occur. "This is a big breakthrough," he said, "but we're not sure it's a panacea."

AI: Mite Is About Right

A panel of senior scientists in the field of Artificial Intelligence (AI) has published what it describes as "a realistic set of goals for the next 20 years." The panelists say they are trying to counter the "persistent pipe dreams and expectations" raised by the government and the news media.

The U.S. Government's Defense Advanced Research Projects Agency (DARPA) has recomended pursuing a goal of reproducing the brain capacity of a bee by 1995. Most AI researchers, however, believe that goal to be hopelessly beyond reach, and many have ridiculed DARPA's recommendations. "A bee is capable of amazingly complex behavior," points out Thomas T.C. Maccarone, the principal author of the new unofficial guidelines. "We think everyone would be better off aiming for something at least vaguely within reach—like a dust mite."

Maccarone says that even a dust mite exhibits some behavior that is too complex for researchers to understand. "But," he says, "a dust mite is a good metaphor for AI workers to think about. Right now AI is more a collection of metaphors than a collection of technologies, and the last thing we all need is some guy at DARPA who has a bee in his bonnet and who wants us to simulate it. So let's all go out and try to make us a dust mite!"

Bugsicles

Insect rights activists, bioethicists, cryogeneticists, and food technology researchers are all pondering the implications of a discovery made inside a small Japanese laboratory. A team of food researchers in Tokyo has successfully revived small insects after deep-freezing them for nearly 2 years. The finding has implications that reach across traditional academic boundaries.

CHAPTER 9

THE SKY AND ITS NEIGHBORS

ASTRONOMY: AN ORIGINAL ABRIDGEMENT
Special Book Excerpt

Howard Zaharoff, Ph.D.
Newton, Massachusetts

From Compressed Science Books: "Condensations Without Condescension"

With pleasure I present my introduction to astronomy, the oldest profession that can be practiced in a vertical position. The televised version will air on PBS in 12 parts.

Early Astronomy

Early ("Primordial") Man viewed the heavens with wonder, for in the starry skies he saw animals, mystical figures, and even an occasional virgin. His successor, Not-So-Early Man, spent less time gazing skyward, thus losing that sense of wonder but not tripping as often. Unfortunately, the rise of civilization enabled the next stage of humankind, On-Time ("Postprandial") Man, to sleep through the night. This development halted progress in astronomy until the invention of insomnia by the Egyptians and Greeks. The cosmology of Egypto-Grecian ("Greco-Egyptian") Man was geocentric, marked by the firm conviction that the center of the solar system was located in the midtown business district.

Later Astronomy

That mistaken belief was refuted in the 16th century by Copernicus, who reasoned that the sun was actually the center. Galileo bravely agreed until the Church

EARLY ASTRONOMY

LATER ASTRONOMY

Jack Tom

threatened to have him charbroiled, when he recanted.

While Galileo was bravely equivocating in Italy, Tycho Brahe was hard at work in Denmark, measuring the positions of stars and losing his nose in a duel. (His daughter, Maiden Form Brahe, coined a famous phrase when she wrote of his tragedy: "He can't see the nose at the end of his face.") On the basis of Brahe's work, Johannes Kepler demonstrated that planets' orbits are ellipses, forever putting to rest the crystalline spheres of Pythagoras and the eccentric epicycles of Ptolemy. Kepler also proposed using the word "satellite" to refer to moons, but this never caught on because it rhymes with neither "June" nor "croon."

Meanwhile, in England, Isaac Newton had discovered gravity. Much of Newton's work was undone in the 20th century by Albert Einstein, whose revolutionary theory of relativity postulated that the speed of light is a constant, that as objects approach light speed their mass increases towards infinity, and thus that astronauts traveling near light speed needn't bother to diet. (For more on 20th century physics, see my fascinating *Whom Did Niels Bohr?*)

What We Now Know About the Universe

The universe was created from a supercompressed cloud of gas, not unlike that caused by cabbage soup. Scientists have fixed its age at about 17 billion years, both by measuring the recession rate of distant quasars and by carbon-dating Jackie Mason jokes. The universe is composed of space, galaxies, and intergalactic dust. Galaxies themselves are composed of space, stars, and interstellar dust. From the omnipresence of dust, we conclude that nature abhors a vacuum and won't pick up a broom, either.

Earth also began as a cloud of gases, and was slowly condensed by gravity into that lopsided ball we call home. Initially, our planet's surface was primordial ooze, out of which came Man, in need of a bath. (For more on our Earth, its neighbors, humankind, and primordial ooze, see my captivating *Atmospheres to Asteroids, Hemispheres to Hemorrhoids.*)

The earth circles the sun along with eight other planets named for Roman deities, from innermost Mercury (named after the God of Analog Thermometers) to outermost Pluto (after the God of Animated Pooches).

Earth has the sun in the morning and the moon at night, a happy and lyrical state of affairs. When the sun is out, the sky is blue, a phenomenon that has puzzled people through the ages. Why, they ask, isn't the sky yellow, or white, or perhaps a nice plaid? The answer is that blue light caroms and careens around more than any other wavelength of light. Hence the tougher question: Why isn't *everything* blue? (For the answer, see my entertaining *Why Isn't Everything Blue?*)

Exploring Space

For jet propulsion, rocketry, even space exploration, we salute the humble "firecracker" and the ancient Chinese who invented it. This discovery led to the development of gunpowder, which was used mainly for warfare until 1957, when the Soviets launched the first unmanned spacecraft. They later became the first to launch a man, which is more economical than a rocket, though it's hell attaching the fins.

But Uncle Sam had the last laugh. Not only was the United States the first nation to put men on the moon, but in 1979 it proved itself foremost with the technologic capability needed to lose a multimillion-dollar communications satellite. (For more on this exciting new field, see my U.S. Department of Publications pamphlet, "A Primer on the Federally Funded Misplacement of Very Big Devices.")

Conclusion

Together we have explored the universe, our solar system, and dust. We have examined Early Astronomy and its successor, Later Astronomy. Yet already humankind has entered the era of "Post-Later Astronomy," a time when rockets travel to distant planets while Earth appears to shrink and people grow bigger. In these dynamic times, people must understand asteroids, Niels Bohr, blueness, and things like that. So buy my books.

HUBBLE TELESCOPE IMAGE OF MARS REVEALS RINGED PLANET

Robert Kirshner
Harvard University, Cambridge, Massachusetts

After months of bad publicity concerning the defective mirror of the Hubble Space Telescope, scientists at the Space Telescope Institute for Spin Control (STISC) are reporting the discovery of a new ring around the planet Mars:

> "We said this was the greatest thing since Galileo, and now we're redoing his work 280 years later," said Jim Pointphail, project historian and head of the Definitive Instrument Resolution Team (DIRT).

Olga Char of the Data Image Restoration Team (DIRT) explained the computer processing used to produce this image from the raw data, "We've had good success with cleaning algorithms in producing rings around many of the objects we point at." Aurora Kirshner, head of the Digital Image Retrieval Test (DIRT), injected a note of caution at a briefing for journalists, "We've had some difficulty with correctly retrieving positions from our archive, but we think we're pretty sure now."

The images were obtained with the Wide-Field and Planetary Camera, which NASA spokesman Ed Whileaway said is used to obtain images of wide fields and planets. "Our mission is 110% in every regard, and we're working to repair it. We believe that management knowhow can turn a minor technical problem into a major feat. There will be no loss of science on my watch," he said.

MARS AS SEEN BY THE HUBBLE SPACE TELESCOPE
The Hubble data support a statement by the esteemed [past] Chairman of the National Space Council, Dan Quayle, comparing Mars with Earth. "Mars is essentially in the same orbit," Quayle said, "somewhat the same distance from the sun, which is very important. We have seen pictures where there are canals, we believe, and water. If there is water, that means there is oxygen. If oxygen, that means we can breathe." In addition to his duties with the National Space Council, Quayle also served as Vice President of the United States. (Photo courtesy of National Aeronautics and Space Administration.)

THE MAN WHO SAVED PLUTO

Steve Nadis
Cambridge, Massachusetts

[*Author's Note:* This story is bigger than you or me; it's about an entire planet. It's also about human nature—greed, squalor, and small-mindedness—and efforts, not entirely aboveboard, to reduce a once-proud satellite to the rank of planetoid. But Pluto was up to the challenge, proving once again to be his old irreducible self.

A word about my friend, Harry the House, who's been the victim of a lot of behind-the-back back-stabbing, illegal eye gouges, noogies, head butts, and cowardly rabbit punches. Not only did he save Pluto, he also saved millions of citizens who were indoctrinated to accept the "fact" that there are nine planets, rather than learning that there are, in actuality, only eight. ("Planet X," of course, is the wild card in his deck, but that's a matter to be addressed later.)

Some malcontents might protest that the aforementioned "H. the H." didn't really "save" anything. They say his victory was Pyrrhic at best and probably only semantic. Nonsense. That line of reasoning is so flimsy as to fall apart under the most casual degree of scrutiny. In fact, it's just plain dumb.]

Let's face it, Pluto was getting a bum rap. They said he was too small to be a planet. Shouldn't even step into the ring with Uranus. The bout with Planet X was up in the air. The astrologists were up in arms. Jimmy the Greek had been canned. The number boys in Vegas were having conniptions. Meanwhile, the press was having a field day. Word had it Pluto would be kicked down to the Asteroid Belt, play a bunch of nowhere towns like Ceres, Pallas, Vesta, Hector, and Herculina. Next they'd be whispering "planetoid" everywhere he went. But what did "they" ever know?

Enter Harry the House, best trainer in the business.

I've known Harry for 20 years, and in all this time, I've never once known him when he wasn't looking for the "best heavyweight prospect in the solar system." Why he gets it into his head that Kid Pluto is such a prospect beats me, but what do I know about the fight game?

Here's the scoop: One night I find myself at the Garden to witness the battle between Frank the Turk and Mack the Knife. I'm there on account of a duckat I copped from a sports scribe at the *Herald*. I tell you about the duckat because I don't want you to think me such a sap as to shell for a ticket with my own scratch, even if I had my own scratch.

At the gate I run into Harry the House, who suggests we hop the shuttle to check out the "best heavyweight prospect in the solar system." I explain that I was just about to step inside for the contest. "Forget the contest," Harry says. "It's just another tank job. Besides, you've got to see this Kid Pluto. He's on the light side— but a real scrapper. Kind of reminds me of Orion the Hunter."

I inform Harry that if he's looking for a backer, he ought to save his breath, as I myself am currently low on that critical green matter. "No problem," he says. "All we need is shuttle fare, and I know you have that. You put the touch on Minnie the Moocher less than an hour ago." There's no use giving Harry the dodge— considering his left hook and reach advantage—although I'm mighty curious as to what stoolie informed him of my current fiscal fluency.

So we trot out to see the prospect. That's no easy task because of his kookie schedule—the graybeards call it a "chaotic orbit"—which makes him one tough bird to track down. But we find him playing rope-a-dope with a one-trick pony called Charon. He'd been running around with her since '78 or before. Calls her his "lunar sparring partner," whatever the heck that means.

Harry signs the Kid for a song and immediately goes to work. The first thing is to get his weight up to where it should be, if for nothing more than to stifle those screaming banshees of gastronomers. Pluto starts bulking up in a big way—pumping some iron, bauxite, and aluminum, and accreting a diet high on methane, dust, and trace metals. "He's small," Harry says, "but solid—a lot denser that anyone gives him credit for. That talk about 'asteroids' really gets me. This kid eats asteroids for breakfast! And his wind is pretty good, too, despite the beef about no atmosphere."

Still, the Kid needs some orbital-vascular conditioning, as would any mug in his whacked-out trajectory. That's where Harry the House—the man who single-handedly stopped Venus's retrograde motion and made her a contender again—enters the picture.

Ask anyone who knows Harry, and they'll tell you the same: He runs the tightest camp this side of the Van Allen Radiation Belt. Rule number one: **NO PRESS.** (He lets me nose around on account of our being pals, and because he knows the odds of my stuff seeing the light of day are as slim as Tyson getting out on good behavior.) Jimmy the Greek had already been sacked for mouthing off about the biggest no-show of the century, Killer (Comet) Kohoutek. Double K had taken a flop, and everyone knew it. Still, the network boys weren't taking any chances, especially with Neptune Day coming up. So the Greek was history. "The last thing we need is to stir up another interplanetary incident," says Harry, a man known for his wily way with the winsome word.

Rule number two: **NO DOLLS.** Fighting and babes are like oil and water, Harry likes to say. They don't mix. Pluto screams bloody murder. You see, he was pretty sweet on this Charon number. Kept him company; added some stability to his life. "The filly stays, or I go," Pluto squawks.

Harry thinks it over and decides to give it a shot. "Besides," he says, "I'm having a hard time finding another mook to dance with the kid." "What's a mook?" I ask. "Like a moon, only with a k," Harry says and winks. Like I says, he's got a willful way with the woeful word.

A month later, Harry's ready to plop Pluto into the ring with a red-hot firebrand named Marcus Mercury (*aka* "Merk the Mark"). Harry gets head commissioner Billy the Fixer on the phone to seal the deal. "No go," says the Fix. "Your boy can't make weight."

"He'll make it," Harry says. "He's been cutting down on the saturated fats, eating a lot of rocks and frozen yogurt. Listen, I don't want no trouble. I don't tell nobody how to run their racket. I just want the kid to get a fair shake." The Fix throws in the deuce. "All right," he says, "just keep that mook out of the ring. I don't want one of those tag-team jobs." "What's a mook?" Harry asks, always the cut-up.

No time for belly laughs; the match was on for Saturday night. I'm no gambler, but I know a good thing when I see one, and this Kid was as fine a gift horse as ever graced the Springs. Sure enough, Pluto knocks off the Mark in four. Then he goes after Uranus and puts more wiggles in his orbit than Planet X ever did. The bout with Planet X—putative dinosaur killer who deep-sixed T. Rex and other notables—was a joke. X went elliptical in less than 30 seconds. He wouldn't be bothering the Sun Belt for another 1,000 years.

And that, sports fans, was how Kid Pluto fought off the Astrology Commission, stayed in the Planetary league, and eventually became "Heavyweight Champion of the Solar System."

CHAPTER 10

FISHES, FROGS, AND THE PACIFIC NAUGA

THE EFFECTS OF ANABOLIC STEROIDS ON NORTHERN GRASS FROGS (*RANA PIPIENS*)

Joseph F. Signorile

Human Performance Laboratory,
University of Miami, Coral Gables, Florida

Introduction

The effects of anabolic steroids on both humans and laboratory animals remain controversial. Some studies have shown steroids to be effective enhancers of muscle mass only when coupled with exercise (Paine, 1986). In other studies, anabolic steroids appeared to act without the intervention of exercise (Dupont, 1967). By far the most striking evidence of the effects of anabolic steroids is seen in women (Mass, 1976). The purpose of this study was to establish the anabolic effects on a nonexercising animal model.

Materials and Methods

Twenty female frogs (*Rana pipiens*) were utilized in the study. Frogs were divided into two groups: a low-androgen nonactivity group (LANA) and a steroid anabolic nonactivity group (SANA). Both the LANA *Rana pipiens* and the SANA *Rana pipiens* were kept on a 12-hour daylight nonactivity/nighttime nonactivity protocol and force-fed a diet of dessicated eel liver and nitrogen-high amino acids. At the end of 6 weeks, animals were examined for changes in a number of variables. The specific variables and the techniques used to measure them are discussed below.

Below: *Photograph of typical SANA frog at the end of the six-week period*

Results

Overall mass evaluation. Frogs were measured using the technique of Perkins (1916). Results are shown in Figure 1.

Jumping distance. The maximum jumping distance attained by each animal following a caudal jab with a soft

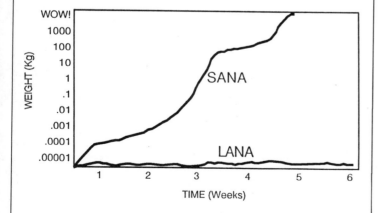

Figure 1. Weight gain of SANA- and LANA-treated frogs over 6-week testing period.

rubber frog prod (Kermit industries #127985-99127) was measured by the linear interpretation method of Leapalunga (1986). Results fell short of expectations.

Strength measurements. Owing to the inability of frogs to grasp implements (Gremlin, 1966), frogs' overall strength was assessed by using their single grasping muscle, the tongue. Frogs were conditioned to lift by being offered an alternative diet of steel-impregnated crickets. Once the animals became accustomed to the taste of the ferric ions, the actual metal dumbbells (Weider Mini-Weights, and Weider Sugar-Frosted Mini-Weights, #10023 and #10023sf, respectively) were readily accepted.

Above: Frog lifting Weider mini-weight, #10023.

Androgenic vs anabolic effects. Typical masculinization of females did occur, often causing severe impairment of normal activity. The most pronounced of these effects was the development of facial and axial hair, as reported in other species (Arugg, 1987). It was necessary to shear frogs before testing.

Frequency analysis of frog ribbits. Since a second test was necessary in order to confirm the masculinization effects of the drug, it was decided to analyze the frequency of the frogs' call (aka "ribbit") to determine whether the animals had the characteristic lowering of the voice associated with males (Jackson, 1987). Fast Fourier analysis indicated a 12,492.3-Hz shift in the mean power frequency of the call due to steroid administration. Results are shown in Figure 2.

Statistical analysis. Data were examined by using MAN-OVA with manipulative measures to correct for outri-

ders and undesirable data points. All data were found to be significant to someone ($P < .00001$). By necessity, post-hop tests were used to examine differences in the strength of leaping evaluations.

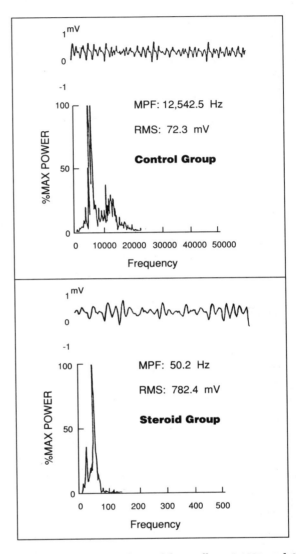

Figure 2. Fast Fourier analysis of frog calls in LANA and SANA frogs.

Discussion

We feel that the results of these experiments speak for themselves.

REFERENCES

Arugg IC. Anabolic steroids and hair growth—A marketing study. *Hair Club For Men Newsl* 17:1–9, 1986.

Jackson M. On the offset of puberty by chemical interactions. *Rolling Stone* 67:12–14, 1987.

Mass Gotalotta. Muscle hypertrophy and facial hair—A covariant model. *East Ger Arch Sports Pharmacol* 27(1):1223–1234, 1976.

Perkins M. You can't tell a frog by its ribbit. *Proc Mut Found Omaha* 13(3):1–61, 1916.

EXPERIMENTAL EVIDENCE AGAINST THE EVOLUTION OF TERRESTRIAL TETRAPODS FROM FISH

Gordon R. Haas

The Ecology Group and Fish Museum,
Department of Zoology
The University of British Columbia
Vancouver, British Columbia, Canada

It is broadly believed that the first terrestrial tetrapods evolved from fish ancestors. However, no direct tests of this hypothesis have been undertaken. Such phylogenetic experiments are usually deemed impossible because they are historic analyses in which the only conceivable tests are inferential. This paper will demonstrate that all of these beliefs are false. The following controlled experiments provide direct tests of just such an historic hypothesis, and the results cast doubt on the evolution of the first terrestrial tetrapods from fish.

Materials and Methods

The laboratory in which I work has approximately two hundred 5-gallon aquaria, twenty-five 10-gallon aquaria, ten 50-gallon aquaria, five 100-gallon aquaria, four fiberglass rearing troughs (50 gallons each), and two sets of four vertical plexiglas holding tanks (25 gallons each). These devices all hold varying numbers and species of fish. The fish are mainly threespine sticklebacks (*Gasterosteus* spp.), but at various times have been or are catfishes (*Ictalurus* spp.), char (*Salvelinus* spp.), lampreys (*Lampetra* spp.), olympic mudminnows (*Novumbra hubbsi*), salmon (*Oncorhynchus* spp.), sculpins (*Cottus* spp.), suckers (*Catostomus* spp.), and various other species of the family Cyprinidae.

These aquaria are roughly equally divided into experimental and control units. The only difference between them is that the experimental aquaria are not covered with clear plexiglas lids, whereas the control aquaria are covered. The fish in the experimental aquaria thus have access to the terrestrial environment via the top of the aquaria; the control fish do not. The laboratory temperature and light regimes correspond to nature. Because the laboratory is unheated, the annual temperature ranges from approximately 0 to 25° C. The natural light regime is maintained by full-light-spectrum fluorescent tubes, controlled by electronic timers.

The vertebrates.

mammals

birds

reptiles

amphibians

fish

terrestrial tetrapods (FEETT)???

fish ——► terrestrial tetrapods (FEETT)??? —≈—► amphibians

The fish are adequately fed, and their aquaria are aerated and filter-cleaned. The aquaria bottoms are covered with sand and/or pea gravel, and some natural vegetation is provided. Dechlorinated water is used.

Since 1984, meticulous records have been kept on the number of voluntary fish experimental evolutions into terrestrial tetrapods (FEETT). A daily survey of the terrestrial environment (laboratory floor) was made for FEETT events, and was supplemented with surveys of all laboratory aquaria whenever possible. The number of fish in each tank was noted, and if one was missing, it was assumed to have undertaken a FEETT and was searched for.

In both the terrestrial and aquatic surveys, the FEETT fish were ultimately classified as a success or failure. "Success" was when a suspected FEETT fish could not be found and was assumed to have developed terrestrial tetrapod characteristics that permitted it to survive and escape detection on land. "Failure" was when a FEETT fish was found in the terrestrial environment and either was deceased or, if still alive, had acquired no terrestrial tetrapod features. Whenever time permitted, recovered FEETT fish were anatomically examined for possible tetrapod characteristics.

Results

I have failed to observe a single successful FEETT event. All FEETT fish were found in the terrestrial laboratory environment, and no fish were ever unaccounted for. This failure has been noted for well over 2,000 sample days and many unfortunate FEETT attempts. This lack of success had no relationship to any of the seasonal or environmental variables monitored, or to aquarium size, fish density, or species. None of the FEETT fish examined had any terrestrial tetrapod characteristics that would have enabled their survival. All FEETT fish came from the experimental aquaria.

Discussion

The lack of any successful FEETT events provides evidence against the broadly held belief that the first terrestrial tetrapods evolved from fish. The evidence for this is particularly powerful here because of the broad range of fish observed and because of the controlled nature of the experiments. The fish observed covered the taxonomic spectrum and include both ancestral and derived members of the grade Pisces.[1] They were a diverse and representative group and were all equally unsuccessful in FEETT events.

The fish were always allowed to attempt to evolve only when they were ready and were never induced to go over the evolutionary aquarium wall. The experimental tanks remained uncovered throughout the entire experiment, and there was always ample food, air, etc., in all the aquaria so that the remaining fish did not die in unusual proportions of unnatural causes. No control fish from covered aquaria attempted a FEETT, and the lids on their tanks were never disturbed by evolutionary activity.

The only terrestrial tetrapod characteristic ever found on the FEETT fish was dry skin. However, my belief about this anatomic observation is that it had actually contributed to the failed FEETT attempt and was not indicative of impending successful terrestrial evolution. These feelings have been further substantiated through confidential discussions with other fish biologists (especially Dr. J.D. McPhail) and with a friend of mine whose father is a dermatologist. These conversations also indicated that my colleagues had never witnessed a successful FEETT. I have also found many unsuccessful FEETT fish in nature, but these of course constitute only circumstantial corroboration. In these latter natural cases, I could not verify whether any fish had been successful in tetrapod evolution and simply went undetected. Nonetheless, no example of a successful FEETT fish has ever been brought to me or to my attention.

This paper also demonstrates that historic analyses can be carried out in a rigorous experimental manner and need not only be inferential. In conclusion, I would like to point out that while my results cast serious doubts on the manner of the first tetrapod evolution, I still strongly believe that evolution does occur. I mention this because I do not want this paper misinterpreted as supporting creationist arguments.

Acknowledgments

I would like to thank everyone for not reviewing this manuscript, anonymously or otherwise. Your contributions to it could not be underestimated.

1. Nelson JS. *Fishes of the World.* New York: John Wiley and Sons, 1984.

THE PACIFIC NAUGA: AN ENDANGERED SPECIES

Robert R. Pascal
Atlanta, Georgia

Little attention has been paid to the threatened extinction of *Cervus nauga naurae*, the Pacific Nauga deer, more simply referred to as the Nauga. For ages, these graceful creatures inhabited several Pacific islands (mainly Naura) in a steady state of survival. The adult female Nauga calves once a year, in the fall, and the young reach sexual maturity in 3 years. Until recently, the deers' only predator was the aboriginal Naura population, whose tribespeople value the hooves of the older adult animals as belt decorations after carving them into deist shapes. For centuries, the Naura tribes maintained strict limits on the number of Naugas that could be killed annually and provided "licenses" to kill the animals as rewards to marriageable males of high standing. The meat of the Nauga was usually preserved and eaten at wedding feasts. It has been estimated that, between 1935 and 1945, there were 1 million to 1.4 million Nauga inhabiting the island of Naura and its immediate neighbors.

The first recognized decline in the Nauga population occurred in 1947, when the birth rate declined to 50% of the expected rate. By the following year, the birth

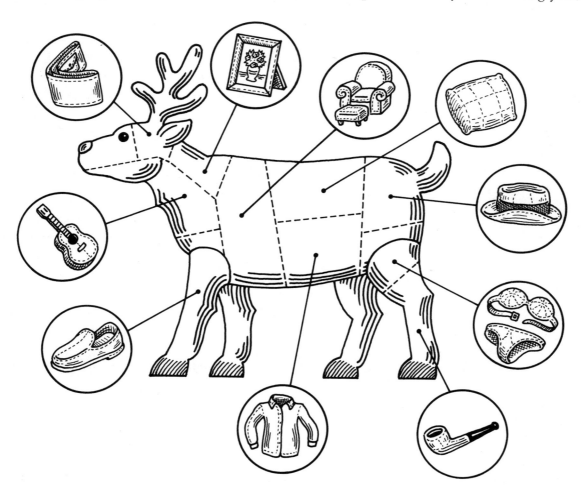

rate had returned to normal. The 1-year decline was ascribed to a large number of *in utero* deaths, possibly due to fatal malformations produced by radioactive fallout from the Bikini atoll nuclear bomb tests. Because of that possibility, 100 Naugas were transported by the U.S. Army to an undisclosed site in the United States for scientific studies.

Although the results of those studies were never published (thus implying that some evidence of radiation damage was, in fact, found), the sacrifice of those 100 animals led to the wholesale slaughter of almost the entire Nauga population. As was common in that period, unwanted, decontaminated animal skins were sold to contractors through the Army's distribution centers. In 1948, one manufacturer purchased 80 available skins, cured them, and introduced them to the public via a major department store. The result was an immediate success for the furniture industry, and a disaster for *Cervus nauga naurae.* Naugahyde was in!

Within a scant 5 years, the demand for Naugahyde chairs, recliners, sofas, ottomans, and chaise-longues exceeded the supply of skins. Traffic in Naugahyde provided vast profits for hunters, shippers, and manufacturers. The methods of killing the Naugas became increasingly brutal. Whereas at first only the adult males were individually taken, the popular demand for the hides brought unscrupulous hunters to the Pacific islands. Traps were set around watering and feeding areas, and Naura tribespeople, with the promise of an increasing supply of Nauga meat, were seduced into dropping their taboos and driving the animals toward the traps. Ensnared animals, regardless of age and sex, were shot and clubbed to death in vast numbers. The skins were removed on the islands and cured in factory ships en route to California. By 1970, only an estimated 10,000 Naugas were alive in the South Pacific, and yet the demand for Naugahyde persisted.

In 1974 a conservation group known as the Espiritu Santo Expedition set out from Seattle to save the Nauga. With 18 members, a three-masted schooner, the "Makepeace," and limited funds provided by conscience-stricken purchasers of Naugahyde furniture and the Southwestern Cattleman's Association, the E.S.E. attempted to persuade the aboriginal Nauras to revert to their socioreligious customs and ban the Nauga hunters from their island. The mission was met with derision. The Nauras had become rapidly, if not totally, Westernized. Nauga meat in every form—including ground patties—had become a dietary staple, and Naura high officials sat on Naugahyde chairs.

Direct action to prevent the slaughter of more Naugas led to the injury of four E.S.E. members and the near-death of one. Finally, on the night of February 18, 1976, 20 mating pairs of Nauga were brought aboard the "Makepeace" and transported to the island of Espiritu Santo under the protection of the New Hebrides Conservation Department.

Today there are no Naugas on the islands of Naura or its immediate neighbors. The demand for Naugahyde is virtually nonexistent as a result of the animal's near extinction and the fickleness of the American consumer. The native population of Naura has returned to its old ways, imposing the old sanctions on the killing of Naugas. Fish is served at wedding feasts, and the old belt decorations of Nauga hoof have become fantastically valuable. In 1984 the island of Espiritu Santo was a haven to slightly more than 600 Naugas.

It is hoped that the Naugahyde fad will never reappear, but equally important is the lesson to be learned regarding other potentially endangered species. The E.S.E. is currently turning its attention to a problem that could duplicate that of the Pacific Nauga. A measurable decrease in the population of the Norwegian Gortex has been observed over the past 2 years, signaling continued vigilance for E.S.E members worldwide.

SEASONAL VARIATION IN NOCTURNAL OCCURRENCE OF *FELIS DOMESTICUS*

Alfred C. Marsh and Betty Jean Cuddles

Institute for the Study of Felicity

For centuries observers have been intrigued and baffled by feline psychology in its many forms. Among the most puzzling problems has been that of nocturnal activity, which appears to vary in location and intensity throughout the year. Early researchers working under primitive conditions apparently experienced difficulty in reliably identifying their subjects.[1] In modern times, the problem was cogently stated by Goose,[2] although we disagree with her conclusions.[3] The most common theory suggests a strong lunar cycle in activity,[4] although a minority believe that fraternization with other organisms is more important.[5]

An obvious flaw in the studies cited above is the lack of quantitative analysis; that is, the reliance upon anecdotal evidence. To remedy this weakness, we have conducted a controlled study over an 18-month period. While we do not deny the existence of a lunar cycle, we have found that a seasonal cycle explains 85% of the total variance in the time-series data set.

Materials and Methods

The experimental setting was a 20' x 20' room containing two dressers, one king-size waterbed, a bookcase, and a radiator of "old-fashioned" (or "hot-water-pipe") variety. The subject was the resident feline, variously referred to in the course of the experiment as "Pussum-tat", "Sweetie-pie,"[6] and "$%!#&%!."[7] As a compromise, the subject will herein be referred to as "The Cat." Participants in the experiment were The Cat (male, neutered) and the investigators, ACM (male, not neutered) and BJC (female, *definitely* female).

Previous pilot observations revealed the existence of four preferred locations of The Cat: (1) between ACM and BJC (on bed); (2) BJC-lateral (also on bed); (3) radiator; (4) outdoors (exact location unknown). These categories were supplemented by a fifth—"Other" (on floor, on bookcase, or under bed)—and listed on data sheets. Observations of The Cat's location were made during the hours of 23.00 and 08.00 daily, with some variation.[8] At least four observations were made each night, and were averaged to produce a single value (see Figure 1 for a typical record). On some nights more or less marked physical activity on the part of ACM and BJC occurred immediately following retiring. During such periods observations of The Cat were neglected in favor of more pressing issues. However, it is our impression that The Cat usually was not present on the bed, although occasional feline attempts were made to participate. Negative reinforcement was provided by ACM in these cases.

	2300	2400	0100	0200	0300	0400	0500	0600	0700	0800
BJC–lateral			✓			✓				✓
Radiator										
Outdoors										
Between ACM and BJC										
Other *Floor: Tripped over in Bathroom*							✓			

April 7, 1984

Figure 1. Sample data sheet for one night of observations.

Results

Nocturnal location throughout the experiment is shown

in Figure 2, plotted as a frequency histogram of nights per month. Although there is some variability in the data, location appears to be nonrandom, and shows significant clustering with respect to season.

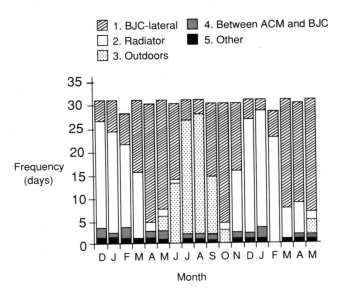

Figure 2. Monthly location of The Cat averaged for each night.

Averaging over the duration of the study, The Cat occupied Location 2 (BJC-lateral) 50% of the time, Location 3 (radiator) 33%, and Location 4 (outdoors) 17%.[9] The distribution is strongly tied to a seasonal cycle, with Location 2 preferred during spring (March-June) and fall (September-November), Location 3 during winter (December-February) and Location 4 in summer (July-August). Due to the nonrandom temporal sampling of our study (an 18-month interval from December of year 1 through May of year 2), the overall mean of our data is biased toward winter and spring. Using a weighted average, we predict that on a typical annual cycle, distribution would be:

       Location 2—50%
       Location 3—25%
       Location 4—25%

Thus, Location 2 remains dominant, but Location 3 emerges as equivalent to Location 4.

Discussion

The strong pattern seasonal variation observed, and tight within-season clustering, suggests the existence of an annual rhythm (cyclicity) in one or more variables. Hor-

mone cycles, often noted in animal behavior patterns, are not relevant here, The Cat having been surgically relieved of such cycles.[10]

While many variables might be invoked, we believe that the location of The Cat can be described adequately as the result of interaction of three behavioral tropisms, each exhibiting a different infradian and/or circannual rhythm. The strongest appears to be thermotropism, which has been widely observed in feline subjects.[11] Thermotropism could account for the dominance of Locations 2 and 3, and the "preferred," though rare, Location 1. There are two sources of heat flux in the experimental setting which seem to elicit a positive thermotropism: BJC and the radiator. The former provides a seasonably invariant but low flux, while the latter is seasonally intermittent, but when functional provides a higher flux than BJC.[12]

Originally, one of us (ACM) maintained that thermotropism alone could explain location preference of The Cat, arguing that the source of high heat flux (radiator) was consistently preferred to the lesser heat source (BJC) whenever the former was functional (i.e., the winter months). Further analysis by BJC, however, has revealed that the radiator is actually functional for a longer period (midfall to midspring) so that, were thermotropism the only factor, seasonal position pref-

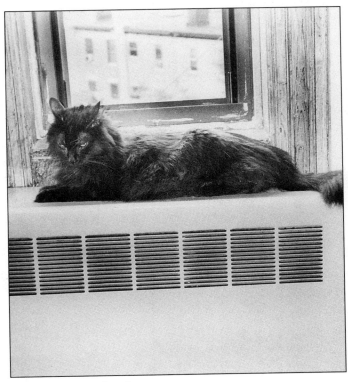

Photo courtesy of Sheilah Scully

erence of The Cat would show a stronger dominance of Location 3. ACM has reluctantly conceded that a secondary drive is present, which he calls "gynotropism;" BJC calls it "love," which is unscientific. This explains the dominance of Location 2, satisfying both drives, which appear to be of equal intensity during seasons of intermediate ambient heat (spring, fall). During winter (low ambient heat), thermotropism dominates and The Cat is located in the area of greatest heat flux (Location 3).[13]

A third and lesser drive appears during summer, when ambient heat is high, so that thermotropism is satisfied in any location. While gynotropism continues to be operational during the day (BJC, personal observation), the third drive, as yet unexplained, comes into play and Location 4 is favored. Since Location 4 cannot be observed from the experimental setting, exact conditions of occurrence are unknown, rendering nature of the drive difficult to determine. Possibilities are "curiosity," "playfulness," and "restlessness."[14] It is also possible that a residual hormone cycle continues to operate.

1. "By night all cats are grey."

2. "Pussycat, pussycat, where have you been?"

3. While it has long been known that "A cat may look at a king," Goose is the only authority to propose a queen as the object of fixation.

4. For example, "The little black cats are maddened/ By the bright green light of the moon"; "Jellicle Cats come out tonight/ Jellicle Cats come one, come all/ The Jellicle Moon is shining bright/ Jellicles, come to the Jellicle Ball."

5. "The owl and the pussycat went to sea," "The gingham dog and the calico cat / Side by side on the table sat."

6. Terms used by BJC.

7. Term used by ACM, not readily translated.

8. For instance, on January 1 of each year, time of observation was from 03.00 to 11.00. Such atypical days were eliminated from the statistical analysis to avoid complicating variables such as photo- and gastrotropism.

9. Location 1 (between ACM and BJC) was probably the most preferred position, but occurred rarely, due to negative reinforcement by ACM.

10. BJC believes that similar cycles are involved in the early-evening activity periods; ACM, on the contrary, contends that hormonal cycles can only be related to nights on which activity did not occur, approximately 4 out of 28.

11. For example: "And now I can sit by the warm fire at the back of the cave for always and always and always."[8] "The cat, if you but warm her tabby skin/The chimney keeps, and sits content within."[9]

12. ACM is probably equivalent to BJC as a source, but test conditions were such as to cancel the effect (see negative reinforcement, above).

13. BJC maintains that this is only because Location 1 is restricted.

14. Terms used by BJC. ACM is still trying to find applicable scientific concepts.

REFERENCES

Cervantes M. *El Ingenioso Hidalgo Don Quixote de la Mancha*. Pt. II, Ch. 33. 1605.

Coatsworth E. The bad kittens, In: *Compass Rose*. 1929.

Eliot TS. The song of the jellicles, In: *Old Possum's Book of Practical Cats*. 1939.

Field E. The gingham dog and the calico cat, In: *Poems of Childhood*. 1949.

Goose M. *Collected Wit and Wisdom of Mother Goose*. 1912 (reprinted).

Heywood J. Proverbs. Pt. II, Ch. 5. 1546.

Kipling R. The cat that walked by himself, In: *Just-So Stories*. 1902.

Lear E. The owl and the pussycat, In: *Nonsense Songs, Stories, Botany and Alphabets*. 1987 (reprinted).

Pope A. *The Wife of Bath. Prologue*. 1.142, 1708.

SAUROPODS AND THE ORIGIN OF FLIGHT

Stephen J. Godfrey
Mississauga, Ontario, Canada

The 19th century discovery of the Upper Jurassic fossil skeleton of *Archaeopteryx lithographica* in the lithographic limestone of Germany marks the second-highest point in the history of vertebrate paleontology and its contribution to the development of evolutionary thought. (The vertebrate paleontologic high point came when we dined on raw blubber from the frozen carcass of an extinct Siberian woolly mammoth at the annual society banquet.) Ever since the discovery of the first specimen of *Archaeopteryx* by Dr. Karl Haberlein in 1861, too much has been written about the ability, or lack thereof, of this early "bird" to fly.

Today, two competing theories on the origin of flight in birds vie for the support of professional paleontologists. The older theory, first proposed by Icarus, suggests that flight came about as the result of frequent jumping and gliding bouts between trees and the ground.

The clawed forelimbs of *Archaeopteryx* were employed to climb into the forest canopy, whence this proto-bird glided from branch to branch and to the ground. The selective advantage favoring the development of enlarged feathers on the forelimb and tail was to reduce the chances of unconsciousness brought about by hitting the ground after an accidental fall off a dew-

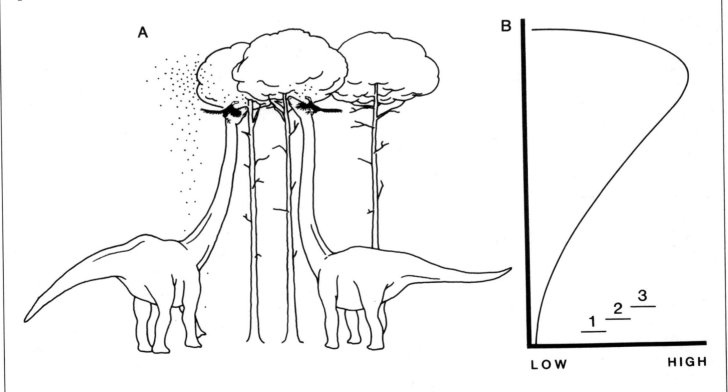

Figure 1A. *Archaeopteryx lithographica* consuming airborne insects (dots) while riding high atop herbivorous sauropod dinosaurs. **B.** Altitude vs insect density plot based on evidence from the fossil record. Insect density is known to have been highest near the upper feeding levels of sauropods. The adaptive value of head-riding is readily appreciated when one considers the numbers 1, 2, and 3 above the horizontal bars, which mark the Jurassic high-jump records for amphibians, reptiles, and mammals, respectively. In this resource-depleted zone a high premium was placed on insectivores getting off the ground.

slick branch at the end of a long night of Bavarian pub-hopping. The alternative theory suggests that *Archaeopteryx* was cursorial, and flapped its proto-wings to escape ground-dwelling predators, to pursue and capture land-locked prey, or to create gigantic dust storms on hot summer afternoons just for the fun of it. The ability to fly came much later as the result of countless catapults from boulder to boulder.

Unfortunately, each camp accuses the other of ignoring evidence when constructing their respective hypotheses. As in other emotional debates, the participants lose the ability to construct alternative theories that may account more parsimoniously for the evidence at hand.

Discussion

After a brief examination of all the factors impinging upon Jurassic life, I now humbly but confidently champion the following internally coherent theory, which superbly accounts for all of the aforementioned evidence: Feathers (and most avian features) evolved in coelurosaurian dinosaurs in order to control the foraging behavior of giant sauropod dinosaurs (Figure 1A). The following comments are intended to convince skeptics.

From the Pennsylvanian to the Triassic eras, most terrestrial quadrupeds were insectivorous. Over 170 million years of selection had produced some pretty stealthy hunters. By the Jurassic era, ground-dwelling insect numbers had dwindled, whereas airborne populations—virtually free from predation—had soared (Figure 1B). The adaptive value of being able to reach these airborne insects was obvious, and the race was on to exploit this untapped food resource. The depleted terrestrial resources pushed some small insectivores up into trees. Others experimented with hard drugs in an attempt to get high.

During the Jurassic era, giant herbivorous sauropods multiplied as they cropped foliage from the forest canopy. At this time a small nomadic herd of coelurosaurian dinosaurs discovered, while practicing an early form of leapfrog, that by racing up the tail, running along the back, and ascending the neck of gargantuan sauropods (for example, *Diplodocus, Aepisaurus, Apatosaurus* (formerly *Brontosaurus*), *Pelorosaurus,* and *Brachiosaurus*), the probability of capturing prey (flying insects) was greatly enhanced. Because there were only a few sensory receptors on the sauropod body, the chicken-sized coelurosaurs went unnoticed. Unfortunately, without some way to direct the movements of the herbivorous sauro-

pods, the nimble coelurosaurs were shuttled far and wide, subject to the whims of these meandering giants. Just which whims most frequently afflicted sauropods is still hotly debated and although it is difficult to be precise at this juncture, it seems self-evident that sauropod head-riding quickly gained favor among small coelurosaurs as the optimum way to maximize insect intake.

As with many evolutionary trends, behavioral novelties initiate structural changes. To their amazement, coelurosaurs discovered that covering the eyes of the sauropod with their forefeet would elicit behavioral changes in the sauropod. Although crude, this method allowed coelurosaurs to direct sauropod movement into insect-rich areas. The degree of behavioral control over sauropods was directly related to the area of the forelimbs that could be devoted to covering the eye while maintaining a grip, so the selective pressure was intense for an increase in the surface area of the arm. Enter feathers.

The totality of adaptive features preserved in the skeleton of *Archaeopteryx* demonstrates that it represents the culmination of this adaptive gestalt. The clawed hands of *Archaeopteryx* were useful in leg and neck climbing and to grip the head of the sauropod. A restoration of the feathered forearm of *Archaeopteryx* placed over the head of *Diplodocus* (Figure 2) demonstrates that the orbital margin could be completely covered. Asymmetrical "flight" feathers developed in the head-riding *Archaeopteryx* because they moved through the air relatively quickly. The fact that skeletal adaptations for powered flight are not seen in *Archaeopteryx* (e.g., no sternum, small scapula and coracoid, etc.) is no longer an embarrassment, since *Archaeopteryx* was not engaging in flapping flight. The shape of the furcula or "wishbone" indicates that the "wings" were not flapping. Nevertheless, the furcula had to resist the transverse tension generated by the pectoral musculature in order to keep the feathered "wings" down over the eyes of the sauropods. Enlarged tailfeathers developed initially as an aid to maintaining balance but were quickly co-opted as a sauropod movement controller. If the tail was bobbed up and down, patting the back of the sauropod's head and neck, *Archaeopteryx* could initiate forward movement in the sauropod (Figure 2). Elongated feathers did not develop on the hindlimbs because there was no adaptive value for them to do so; through a complex series of partial or complete eye-covering behaviors combined with tail bobs, *Archaeopteryx* could

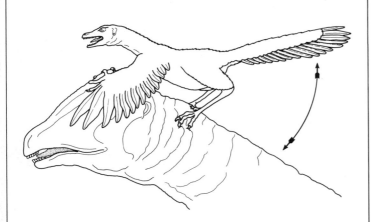

Figure 2. *Archaeopteryx lithographica* in a life-like stance on the head of the sauropod *Diplodocus*. The feathered arms completely cover the orbits. Movement in any direction could be initiated and controlled by combining "wing" position with tail-bobbing.

manipulate the largest of sauropods into insect swarms.

The feather eye coverings also had a calming effect on the sauropods, much as blinders do when employed on carriage horses. Adult sauropods were so large that they had no reason to fear attack from marauding bands of relatively puny therapods (carnivorous dinosaurs). In spite of the fact that the sauropods had no reason to fear theropods, their skittish temperament interfered with foraging. When inhaled, feather dust, with its mild analgesic properties, calmed sauropod jitters. The pos-

terodorsal placement of the external nostrils in some sauropods has defied explanation for over 100 years. Because head-riding was beneficial to the foraging behavior of sauropods, there was a selective pressure that favored the posterodorsal migration of the external nostrils because their pungent breath attracted flying insects, which were readily consumed by *Archaeopteryx*. The poor sense of smell in living birds (dead birds smell quite strongly) is a result of the close proximity of *Archaeopteryx* and its kin to sauropod breath.

These feathered proto-birds quickly monopolized available sauropods, displacing closely related but less completely feathered coelurosaurs. *Archaeopteryx* species at the lower end of the pecking order and those with spare time tried to manipulate other dinosaurs. Although most could be mastered, none proved to be as beneficial as the sauropods. Ceratopsians (or horned dinosaurs, like *Triceratops*) were the most difficult to head-ride, and head-riding pachycephalosaurian dinosaurs, with their head-butting antics, were clearly nonadaptive (Figure 3).

Conclusions

Sauropod head-riding inadvertently facilitated the development of skeletal features required for powered flight. However, the prototypes were selected in order to improve the directional control exercised by the coelurosaurs over sauropods.

Figure 3. The head-bashing behavior of pachycephalosaurian dinosaurs proved to be nonadaptive for some *Archaeopteryx* individuals.

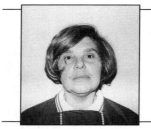

Styles, trends, and tidbits culled from leading research journals
by Alice Shirell Kaswell

The number 17 has received a remarkable amount of attention from scientists in virtually every field. (See, for example, Lee Mondshein's "The Significance of the Number 17 to Mathematicians of My Generation" on page 71 and Dudley Herschbach's "Sweet Seventeen" on page 72.) Regular readers of this column are aware that the publication of record for research related to the number 17 is the eponymous research journal ***Seventeen***. The March, 1991 issue of ***Seventeen*** presents findings of great interest.

Skin as an Ingredient
Reports on pages 2 and 19 of ***Seventeen*** present chemical analyses of Noxzema skin cream. The principal finding is that, together with other ingredients, healthy skin is in Noxzema. There is a discussion on pp. 16–17 concerning the case of a young woman whose eyelashes are a mile long. A groundbreaking applied mathematics paper on p. 22 outlines an algorithmic approach to color-coded curl care. The paper also demonstrates that extra-body formulas add volume.

Superiority and Zum Zum
A report on p. 38 of ***Seventeen*** explains that Neutrogena Shampoo has always demonstrated consistently superior rinsability. Investigator Howard McLaren of Bumble & Bumble trimmed a volunteer subject's hair to her earlobes (McLaren's methodology is detailed on p. 64). The subject reported that, as a result, she washes her hair and rushes out the door. Authentic Paul Mitchell products have been tested and found to be completely

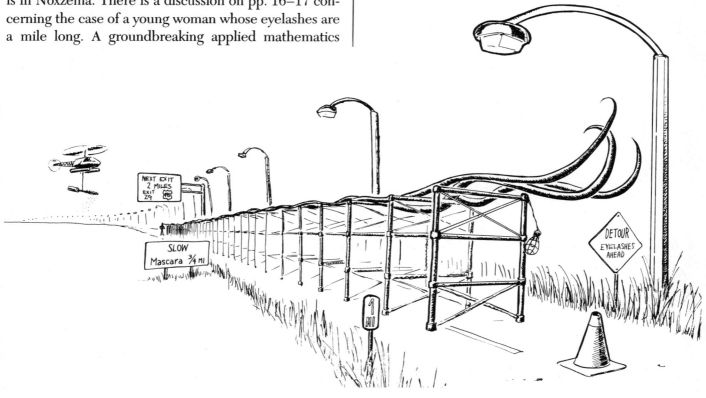

free of cruelty, according to the data on p. 74. Jolen is to be used anywhere one needs paling. The details are presented on p. 80. A controversial subject, Zum Zum, is stripped to its essentials in a report on pp. 89–91.

Humectress Drenching, Papaya, and the Moon

A clinical investigation described on p. 105 of **Seventeen** finds that one should drench beautiful hair with Humectress. A panel of researchers discuss (pp. 112–113) what their fathers taught them about creating shampoo formulas. The lessons were indirectly responsible for the recent development of the well-known environmental formula Aussie Mega Shampoo With Papaya. A report on pp. 122–123 explains that Dep Shampoo Plus Conditioner is the only all-in-one shampoo with natural botanical extracts. It also explains that Dep has no causal relationship with skin that is too blotchy, too sensitive, too oily, too lizardy, too large-pored, too olive, too dry, or just too much like those NASA photographs of the moon's surface.

Photo courtesy of N.A.S.A.

Bongo 5, 7, 9

Seventeen presents a warning (p. 140) that Citre Shine (and in particular, Shine Miracle) has not been tested on animals. In the absence of such tests, the substance should be used only on (or in) humans. A scientific proof (p. 217) reveals that active protein is absorbed right into your hair. The proof is subject to a money-back guarantee. **Seventeen** also has a special report (p. 227) about a phenomenon known as "Bongo 5, 7, 9."

Smoke and Collisions

Vol. 43, no. 1 of the research journal **Modern Bride** presents a number of important findings. A report on p. 6 describes an ultralight cigarette that gives off 70% less smoke from the lit end, under laboratory conditions. No data are presented concerning the other end, and no data are presented concerning the middle, but research assistant Pamela Dennis is credited with authorship of the fashions. A kinetics investigation on p. 30 finds that tables make a big impact.

Brides and Brains

Modern Bride also presents a psychology report (p. 78) outlining the results of a new intelligence survey: There is a high correlation between a bride's intelligence and her demonstrated level of trust in the JC Penney Bridal Catalog. A related report (p. 141), concerning the relationship between the inner ear mechanism and higher cortical functions, shows a correlation between a bride's sense of balance and the cognitive underpinnings of her flatware. **Modern Bride** presents a striking interview (pp. 206–230) with investigator Bobbi Brown. Brown has discovered that one cannot change the shape of one's face with makeup. Brown recommends adding interest at the top of the head; she does not specify a mechanism for doing so.

Vocal Flora, Malfunctioning Rubble

Investigator Barbara Milo Ohrbach's seminal linguistics work is discussed on p. 440. Ohrbach has analyzed the language of flowers. **Modern Bride** also reports (p. 482) that the benefits of being massaged with all-natural essential oils are praised by scientists from all nations. Investigators Anthony Pietropinto, M.D., and Jacqueline Simenauer explain in detail that without that blueprint we call "desire," sex can be a heap of malfunctioning rubble.

Mental Garlic

The research journal the *New York Times* (March 17, 1991, vol. CXL, no. 48,542, part 2) assesses the prospects for future funding of basic research. An analysis (pp. 28 and 73) by Laurel Graeber quotes analyst Andres Shore: "Skin care isn't there yet. But cologne is relatively innocuous." Graeber presents the case for faux granite in bottle design, citing the cases of Fendi Uomo and Red for Men, a new Georgio Beverly Hills substance that contains 551 ingredients. A first-person history of science report (pp. 50–72) by investigator Michael Lindsay-Hogg describes fellow researcher Jeremy Irons's response to a standard psychological test involving the performance of simultaneous mental tasks. Lindsay-Hogg asked Irons to describe the first time they met: "Jeremy Irons looked thoughtful, either trying to recall or trying to separate the garlic from the chili in his spaghetti."

Such Is Man

New York Times investigator Liz Claiborne (p. 73) has been conducting tests with her 1½-year-old scent for men, Claiborne. Claiborne's investigations led her to conclude that "Man is not so simple, after all."

Deconstructing Blonsky

Claiborne's finding is confirmed by another report in the same issue of the *New York Times*. Investigator Marshall Blonsky, a deconstructionist semiotician and professor, has been traveling around the world for the last 4 years interviewing the cultural figures who create the myths that ignite our desire and make us who we are. His interdisciplinary survey (p. 74) probes researcher Gabriella Forte's unsuccessful experiment in which she dressed Blonsky in double-pleated baggy pants that felt drafty and did not inform Blonsky that he was supposed to be one of fashion's new penitents. Gianfranco Ferre's people in New York dressed Blonsky in a chocolate-brown leather greatcoat that came down to his knees. Rosita Missoni wanted to colorize every inch of Blonsky at her factory in Sumirago, Italy. Alexander Julian constructed another Blonsky. The second Blonsky looks intellectual; Julian used herringbone to give the second Blonsky a professional injection.

Inhaling Moth Balls

Sometimes, an investigator's discovery can seem entirely contrary to previous experience; in such cases, the scientific community may be reluctant to accept the discovery. Investigator K. Wayne Wride, writing on page 2 of the research journal *Organic Gardening* (vol. 38, no. 3, March 1991), has made such a discovery and is encountering skepticism. "When I tell you that ice cream can be frozen and dispensed in about 60 seconds," he writes, "you will not believe me." Investigator Marilyn Dupont reviews (p. 7) recent studies that have shown that cats who inhale moth balls or crystals on a regular basis may develop serious health problems. Dupont says that the health problems identified in the studies were either kidney or liver problems, and that she has forgotten which.

In re: Bat Guano

Organic Gardening investigator Thomas Arndt says (p. 10) that he is very pleased with the results he gets from bat guano. In a report on the same page, investigator J.E. Williams speaks out for a neglected nut. Finally, investigator Jeff Cox relates the case (p. 104) of a woman who instructed 10-year-old boys to beat a potato bug to death with a baseball bat.

CHAPTER 11

BIOLOGICAL INSIGHTS

AGING: A CONTAGIOUS DISEASE

John A. Robbins and Jochen Schacht
Ann Arbor, Michigan

It has traditionally been assumed that aging is an inevitable process associated with life. Only recently has research been directed toward elucidating the molecular nature of this phenomenon. Of the two major hypotheses, the first postulates life-span as genetically predetermined. The second considers aging to be a result of negative environmental influences on cellular metabolism and physiology.[1] No attention has yet been paid to the possibility of aging as a contagious disease.

This possibility is suggested by a number of frequently overlooked facts. From birth, we are surrounded by an aging population that can pass the disease along. Babies show little evidence of aging, but after a few years of exposure to the adult environment the aging process becomes manifest. Finally, similarities between aging and slow virus diseases are striking:

1. Both show late onset of disease.
2. The viral nature of the disease is difficult to diagnose.
3. Both have lethal outcome.

To test the hypothesis that aging might be a contagious disease, we conducted a 6-year epidemiologic study of aging in two environments: first, in an environment with a high concentration of young people, and second, in an environment with a more seasoned population. If our hypothesis is true, differences in aging should be found under these conditions.

Methods and Results

The study was conducted during 1974–1980. For an environment rich in young people, three local nursery schools were selected. Initially, the population was asked to fill out questionnaires about their age. This procedure was not satisfactory; 47% of questionnaires were not returned, and 97% of those returned were mutilated or had been used for a multitude of purposes rather than the intended one. We then resorted to extracting the age of the students from school records. For the second, apparently adult, group, we selected the authors' two laboratories, where questionnaires were distributed.

The careful design of the study assures its validity: The same three nursery schools and the same two laboratories were polled over the 6-year period. The results are summarized in Table 1.

Discussion

It is generally assumed that the chronologic age of a

Table 1. Influence of Real-Time on Aging in Child and Adult Environment.

	1974	1976	1978	1980	Net increase per calendar year
Child environment[*]	3.9±0.6	4.1±0.7	3.8±0.7	3.9±0.6	±0.0[†]
Adult environment[**]	29.5±6.7	31.5±6.7	33.3±6.8	35.1±6.9	±0.9

[*] Average age ± SD in same 3 local nursery schools over period indicated. n = 63 per year.
[**] Average age ± SD in same two (authors') laboratories with same staff members over period indicated. n = 11 per year.
[†] Difference between rate of aging in the two environments is significant at the $P \ll 0.001$ level.

person increases at the same rate as real time. Indeed, this was found to be almost correct for the adult environment in our study. Even here there was no linear progression with time, since one staff member's age was listed as 29 in each of the years 1976, 1978, and 1980. This phenomenon of nonaging in an adult environment requires further investigation. However, the salient finding of our study is that there was no net increase in the age of the population in the young environment. This difference of aging in the two groups is statistically highly significant and clearly indicates that aging is absent from a young environment.

The inevitable conclusion from this study is that aging occurs as a consequence of prolonged exposure to an adult environment. It is not known at this point whether the contagious agent is of viral or bacterial origin, although, as was pointed out above, a viral origin seems likely.

Acknowledgments

The authors wish to thank Helga Schacht for collecting the nursery school data. This work is not yet supported by research grants from the National Institute of Health.

1. Hocman G. "Biochemistry of Aging." *Int J Biochem* 1979; 10: 867–876.

NO-Acetol

Dear Sir:
We should like to report the following contraceptive compound which was recently almost synthesized in our laboratory.

The compound is NO-Acetol and its structural formula is shown herewith. Please note that this is a highly charged molecule causing some distortions.
Truly yours, etc.
X. Perry Mental
Institute of Biodynamics

Dear Sir:
I have been pursuing a line of research which included, however unwanted, the contraceptive NO-Acetol as an active ingredient. The results of my experimentation have shown conclusively that alcohol, when applied to the user of NO-Acetol and when accompanied by a change in heat, rearranges its structure to yield the situation shown below. There are aspects of this reaction, however, that defy natural laws and standards of physics: (1) The conservation of energy is never observed.

(2) The curious "blending of the bonds" in the S-like configuration of carbon and oxygen has never been known to exist. (3) The resultant compound is rather short-lived, unless the situation is stabilized by the addition of a ring of gold or silver atoms.

$$N_3^{++}O_6 + CH_3CH_2OH \xrightarrow{\Delta H} NH_4^+ + CN_2O_5H_2 + CO_2$$

OR:

Author's note: In this structure, -NO- is present in only two positions.

Sincerely,
Christopher H. Price
Coronado, California

LOGARITHMIC AND ARHYTHMIC EXPRESSION OF A PHYSIOLOGICAL FUNCTION (ω)

R. Arnold Le Win

Trou du Bois

Introduction
Materials and Methods
Results
Conclusions
Discussion
Summary*

---

*The actual paper is not reproduced here because, as the author explains in his letter to the Editors: "This paper was submitted to the **Journal of Biochemical and Biophysical Choreography,** the Editors of which advised me that the illustrations were irreproducible for technical and other reasons. The Editors of the **Journal of Animal Misbehaviour and Biophysics** suggested that the section *Materials and Methods* was unnecessary, in view of the peculiar nature of the paper. On the suggestion of the Editors of the **Journal of Comparative Eulogy and Biophysics,** I have omitted the *Results* which were considered inconclusive and *Conclusions* which they felt were unjustifiable. The **Journal of Dynamic Penology and Biophysics** recommended deletion of the *Introduction,* which was circuitous and gratuitous beyond reasonable limits, and the *Discussion,* which appeared incorrigible. Finally, the Editor of **Emetic and Enuretica Acta** suggested that the *Summary* be reduced to 3% of the total length of the text. These suggestions have been adopted."

REFERENCES

Ben Lamond M, Morowitz A, Horowitz B, Tomorrowitz G. The path of pi-mesons in an oscillating piezo-electric field in Israel. *Pal J Chem Phys* 1954; 17:81–88.

Itsu, Mitsu, Tauhyahara, Isochie, Shugiwary, Hatsuyama, Miwa, Fan Me Pink. Electron microscopy and fine structure of the limiting membrane of the flagellar mid-piece of the sperm of the gutter-urchin, *Unclepsammechinus militaris. Exper Cell Res* 1949; 4:18–20.

Jones J. The effect of 2, 4:D., DNP, IGY, and maleic hydrazide on the growth of excised spikeley initials of "Early Glory" oats. *Botan Gaz* 1950; 66:14–61.

Juan Don, Smith Phyllis, Hatsui Irene, Fullmann Hermione, Wysckawa Jane, Blz Bella, and Oginski Thelma. Traptocachuamycin, a new antibiotic. *J Biol Chem* 1951; 67:1056–1066.

Melville H. The path of nitrogen. XLVIII. N^{15}_2-fixation and $N^{15}H_2CO.N^{14}H_2$ production in Chilean onions. *Arch Biochem Biophys* 1956; 40:15–18.

Melville H, Washington G, Lincoln A, Cadillac de V. The path of nitrogen. CXL. Absence of demonstrable $N^{15}H_2CO.N^{14}H_2$ production and N^{15}_2-fixation in Spanish onions. *Arch Biochem Ciophys* 1957; 41:156–160.

Pearlz P, Schwein Alicia. A new bioassay for pseudo-isocobalamin B_{2a}, *Ps. pseudosocobalaminovorans* n sp. *J Bact* 1945; 16:280–281.

Ramakrishnamaswami Krichnamaswamirama. Curvature of high-frequency ultrasound waves in distilled Indian water. *Proc Ind Acad Sci* 1951; 18(B):1–243.

Schitz K, Spitz G. Urea excretion, growth hormone production, and caudal temperature of the 6-week-old hypophysectomized, adrenaloctomized, tonsillectomized castrated albino hamster. *Proc Soc Exp Biol Med* 1956; 50:2–4.

Shadrach C, Meschach H, Abednego H, Abednego C. An anaerobic heat-resistant monglaellate ornithine producing sulfur non-purple bacterium isolated from the rectum of the goat. *J Bact* 1944; 70:1–11.

Smith AK St G, Smith Esther St G. Phenylphytalylthiophtalyl chloride, a new reagent for nonoses in paper chromatography. *Biochem J* 1944; 71:90–93.

Spitz G, Schitz K. Urea excretion, caudal temperature and growth hormone production in the 5-week-old hypophysectomized, adrenaloctomized, tonsillectomized castrated albino hamster. *Proc Soc Exp Biol Med* 1956; 50:4–5.

Strickstraw A. The fate of cats. 27. Glycero-1,4-*alpha*-felitol, a new lipid component of the milk of the lion, *Felis leo. Biochem J* 1946; 73:108–113.

Vääräähääha Willi, Søderhør G, Torenssen A, Jonsson, The. Fertility, isotonicity, and agility of the sperm of the gutter urchin *Unolopsammechinus militaris Exper Cell Res* 1949; 4:21–30.

Winken W, Blinken B, Nodd AHH. ATP, ITP, and TVA. *J Biol Chem* 1951; 69:12–18.

Wotherspoon Jane-Marie. Acetyl-N-iso-hydroxybutyl-ethanolaminyl anisole esters in eels (with statistical appendix by the Rt. Hon. Earl of Wakehampton). *Biochem J* 1944; 71:94–130.

Zwickoloff LK, Arnail de la Foret JP X. The effect of maleic hydraide, 2, 4:D coconut milk, and gunpowder on germination, internode length, flowering, root production, and titanium uptake in the Biloxi soybean. *Botan Gaz* 1953; 69:100–131.

(Editor's remarks: Since none of these references could be verified, the manuscript was sent back to the author with the suggestion to revise the references.)

BUTTOCK-DIMPLING IN MIDDLE-AGED WOMEN: A CROSS-CULTURAL STUDY

Marc A. Olshan
Alfred University,
Alfred, New York

The buttocks often play an important underlying role in human relations. They are justly celebrated in art and literature. For example, they figure largely in the work of Rubens and Renoir and are behind much of the social commentary of Lawrence and Updike. As a subject of scholarly inquiry, however, they remain for the most part unembraced by social scientists. This research[1] represents a modest attempt to buttress what is, at bottom, a paucity of reliable information.

Methods and Results

The study was undertaken at two North American research sites (in Canada and Mexico), thus controlling for latitude. Samples of women aged 30 to 50 were selected using the reputational method. Data were collected from each subject for five variables: dimple density (units/cm^2), dimple diameter (mm), dimple depth (mm), age at onset of dimpling (no. of years), and marital status ($+/-$).

Hindsight makes clear that other variables such as dimple dispersion and dimple duration might have been profitably included. However, the time-consuming, not to say ticklish, nature of the data collection process precluded a more global approach. Another methodologic problem concerned buttock configuration. The researcher is confronted with a staggering array of shapes, the import of which is intuitively clear. But quantification of this complex reality remains elusive. Further research is imperative.

A two-tailed probability test detected no meaningful difference between the Canadian and Mexican samples, so they were merged into a single data set (N = 862). These data were then subjected to analysis using several packaged statistical programs in a search for significant deep-seated correlations. Results are shown in Table 1.

Table 1. Correlation Coefficients for Variables Analyzed.

Variable	1	2	3	4	5
1. Dimple density	1.00				
2. Dimple diameter	.27	1.00			
3. Dimple depth	.31	.15	1.00		
4. Age at onset	.19	−.08	.23	1.00	
5. Marital status (+)	.87	.66	.74	.58	1.00

Discussion

The high correlation between marital status and the various dimpling indicators makes clear the aesthetic value that men place on adequate dimpling in women. Unmarried women were considerably less dimpled than their wedded counterparts. Critics will no doubt assail this inquiry for its reliance on soft data and use of *a posteriori* logic. But (and this is a big but) both are dictated by the pioneering nature of the research.

Acknowledgments

To the handful of women who first stimulated his interest in this topic, the author extends thanks from the bottom of his heart.

1. See also Epps. *Buttock-Dimpling in Middle-Aged Men: A Cross-Cultural Study.*

REFERENCES

Dillstone LR. Buttock dimpling in literature: A bibliographic peek. *Int J Appl Anat* 1976; 7:914–945.

Epps G. Buttock-dimpling in middle-aged men: A cross-cultural study. *Stern Inst Series Soc Physiol* 1989; #42.

Olshan MA. Buttock-dimpling in neonates. *J Ped Groups,* in press.

THE AH GENE: IMPLICATIONS FOR GENETIC COUNSELING

Lawrence L. Kupper

The University of North Carolina, Chapel Hill, North Carolina

Introduction

It is generally agreed that the most common chronic diseases have both environmental and genetic causal components. In recent years a considerable amount of genetic research has been directed at identifying specific genes that place people at high risk for developing such diseases. It is the purpose of this paper to discuss evidence supporting the existence of a gene (henceforth called the AH gene) that predisposes an individual to chronic behavior in an obnoxious, boorish, selfish, overbearing, and generally offensive manner. In our terminology, such an individual will be said to be acting like an AH.

Recent research (Rude, 1988) established that the percentage of adults in the United States exhibiting chronic AH behavior is about 32% (95% confidence interval, 27-37%). The negative impact on our society of such a large number of AHs cannot be overempha-

sized. Clearly, the demonstration of a genetic (or hereditary) link with AH behavior could have strong implications with regard to genetic counseling. These and related societal issues will be expanded upon in the discussion section of this paper.

Genetic Theory and Proposed Research Methods

Following classical genetic theory, I postulate the existence of four alleles (denoted A, a, H, and h), which I henceforth refer to as *rectalleles*, for obvious reasons. Each pair of rectalleles constitutes a genotype; with four alleles there are 10 possible genotypes (disregarding allele order). An individual carrying the AH genotype will be referred to as a "complete AH," while an individual carrying either the Ah genotype or the aH genotype will be referred to as a "half AH." Noting that the total physical expression of a genotype is called its phenotype, I further postulate that being a carrier of one of these three genotypes is strongly associated with exhibiting chronic AH behavior (ie, with being phenotypically a "real AH"). It is to be expected that an AH-

Figure 1

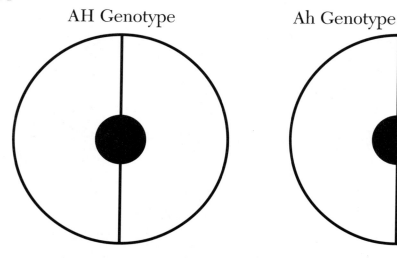

AH Genotype   Ah Genotype   aH Genotype

carrier will, on average, exhibit even more observable characteristics of an AH than will either Ah or aH carriers. Carriers of the other seven possible genotypes are assumed to produce "AH-clean" phenotypes. The AH, Ah, and aH genotypes are pictures in Figure 1.

Epidemiologic evidence (Anass, 1989) suggests that AHs tend to cluster within families more than would be expected on the basis of environmental influences alone. Hence linkage analysis (Ott, 1985), a particular type of pedigree analysis that focuses on intrafamilial patterns of illness, would be ideally suited for establishing the existence of the AH gene.

The methods of linkage should work quite well in studying the families of specific subgroups for which the AH gene is expected to be highly prevalent.[1]

Discussion

As indicated above, strong epidemiologic evidence supports the existence of the AH gene. We would all agree that our lives are adversely affected each and every day by AHs from all walks of life (from department store clerks to high-level political officials). Indeed, it can be argued that almost all of the world's problems are due to some degree to the influence of the AH gene. Hence I strongly recommend that future genetic research efforts be directed at establishing the existence, and determining (e.g., via blood sample DNA analysis) the

chromosomal location of the AH gene. In addition to linkage analysis the molecular genetic-based methods of Swift, Kupper, and Chase (1990) would be directly applicable.

Once the existence and location of the AH gene have been determined, couples with at least one member identified as an AH should receive genetic counseling to caution them that they are at high risk of producing yet another AH. If such couples are still willing to take such a risk, then they should be encouraged to obtain long-term family counseling to minimize environmental influences that predispose to AH behavior.

Acknowledgments

The author wishes to acknowledge helpful discussions with Mr. Thomas D. Higgins III and Mr. Barry H. Jaeger about legal, social, political, and moral issues regarding the AH gene.

---

1. Examples of such subgroups include politicians, physicians, lawyers, used car salesmen, and academicians, to name a few.

REFERENCES

1. Anass RU. Do uncouth people cluster together like birds of a feather?. *Ann Epidemiol* 1989; 8:42–50.

2. Ott J. *Analysis of Human Genetic Linkage.* Baltimore, MD: Johns Hopkins Univ Press, 1985.

3. Rude IM Estimating the prevalence of impolite adults in the USA. *J Obnox Behav* 1988; 69:812–822.

4. Swift M, Kupper LL, Chase CL. Effective testing of gene-disease associations. *Amer J Hum Genet* 1990; 47:266–274.

CHAPTER 12

HAIR AND UNIVERSAL UNDERSTANDING

FELINE REACTIONS TO BEARDED MEN

Catherine Maloney,[a] **Sarah J. Lichtblau,**[b] **Nadya Karpook,**[c] **Carolyn Chou,**[d] **Anthony Arena-DeRosa**[e]

From [a]*Fairfield University, Fairfield, Connecticut;* [b]*University of Illinois, Champaign, Illinois;* [c]*University of Florida, Gainesville, Florida;* [d]*University of Pennsylvania, Philadelphia, Pennsylvania;* [e]*Harvard University, Cambridge, Massachusetts.*

Abstract

Cats were exposed to photographs of bearded men. The beards were of various sizes, shapes, and styles. The cats' responses were recorded and analyzed.

Findings of Prior Investigators

Boone (1958) found inconclusive results in studying feline reactions to clean-shaven men. O'Connor and Brynner (1990) found inconclusive results in studying feline reactions to shaven heads. Quant (1965) found inconclusive results in studying feline reactions to bangs. Seuss (1955) found inconclusive results in studying feline reactions to hats. Ciccone (1986) found inconclusive results in studying feline reactions to hairy legs. Other related studies (Smith/Brothers 1972, Conroy 1987, Schwarzenegger 1983) have since been retracted because the investigators were not able to reproduce their results.

Norquist (1988) performed a series of experiments in which cats were exposed to photographs of Robert Bork[1] (not pictured here), a man whose beard is confined largely to the underside of the jaw. After viewing the Bork photograph, 26% of the cats exhibited paralysis of the legs and body, including the neck. An additional 31% of the cats exposed to the Bork photograph showed other types of severe neurologic and/or pulmocardiac distress, and/or exhibited extremely violent behavior. Because of this, we did not include a photograph of this type of bearded man in our study.

Materials

Five photographs were used in the study. The photographs, reproduced here (Figure 1), display a range of different types of bearded men. (As noted above, one

Figure 1. Left to right: Bearded man JDR-14: John Daniel Runkel, MIT president 1870–1888. Full white beard with central "inverted V" part. Bearded man JMC-3: James Mason Crafts, MIT president 1897–1899. Ripened white "bush whiskers" beard with clean-shaven chin. Bearded man 4/4: Constant Désire Despradelle, Rotch Professor of Architecture at MIT 1893–1912. Full dark semicircular beard. Bearded man HSP-2A: Henry Smith Pritchett, MIT president 1900–1907. Trimmed dark/gray beard. Bearded man NW-736: Norbert Wiener, MIT Institute Professor 1959–1964. Gray triangular beard, cybernetic.

type of bearded man was, however, excluded from use in this study.)

The test subjects were female cats, all between the ages of 4 and 6 years. A total of 214 cats participated in the study. Three cats died during the study, due to causes unrelated to the bearded men. In addition, 15 cats gave birth while viewing the photographs.[2]

Methods

Each cat was exposed to the photographs. One photograph was shown at a time. Each photograph was visible for a span of 20 seconds. The photographs were presented in the same order to each cat.

Above: A feline subject reacts to a photograph of a man with a full dark semicircular beard.

Cats were held by a laboratory assistant while viewing the photographs. To ensure that the cats were not influenced by stroking or other unconscious cues from the assistant, the assistant was anesthetized prior to each session. The cats' reactions were assessed for changes in pulse rate, respiration, eye dilation, fur shed rate, and qualitative behavior.

Results

Quantitative results are average values calculated over the entire feline subject population (Table 1); qualitative results (Table 2) are broken out by percentages of the subject population.

Interpretation

1. Cats do not like men with long beards, especially long dark beards.
2. Cats are indifferent to men with shorter beards.
3. Cats are confused and/or disturbed by men with beards that are incomplete (eg, Bork) and to a lesser degree by men whose beards have missing parts (eg, Crafts).

These interpretations are not categorical. They are subject to several obvious qualifications. The most notable are listed below.

Qualification A. This study excluded photographs of men with beards confined largely to the underside of the jaw (see discussion above of Robert Bork). While data are available from studies conducted by other investigators, those studies made use of a different methodology than the one we used in our study. We are therefore hesitant to interpret our findings in light of the "Bork" findings, or vice versa.

Qualification B. This study was conducted with photographs of bearded men. In a future study we intend to investigate feline responses to animate bearded men. A large number of factors might produce significantly different results in the two studies. In particular, there has been speculation that bearded men produce pheromones which could have a significant effect on cats.[3]

Table 1. Quantitative Results.

	Pulse rate	**Respiration**	**Eye dilation**	**Fur shed rate**
Runkel (Figure 1)	+42%	+186%	+23%	+12%
Crafts (Figure 2)	Unchanged	Unchanged	+1%	Unchanged
Despradelle (Figure 3)	+87%	+317%	+31%	+19%
Pritchett (Figure 4)	+2%	+3%	+3%	+2%
Weiner (Figure 5)	Unchanged	Unchanged	Unchanged	Unchanged

Table 2 Qualitative Results

	Attack photo	Flight	Lick photo	No visible response
Runkel (Figure 1)	52%*	34%	—	14%
Crafts (Figure 2)	2%	1%	1%	94%
Despradelle (Figure 3)	79%†	19%	—	2%
Pritchett (Figure 4)	7%	1%	—	91%
Weiner (Figure 5)	—	—	—	100%

*Attacks included hissing, spitting, and generally agitated behavior.
†Attacks included hissing, spitting, generally violent and agitated behavior, chaotic tail twitch, screeching, and incontinence.

Acknowledgments

The author wishes to thank The MIT Museum for allowing us to use photographs from its Bearded Men Collection and for generously granting permission to reproduce the photographs as part of this research report. Special thanks to Sally Beddow for assistance in selecting appropriate photographs (the Collection includes more than 71,000 photographs of bearded men) and to Warren Seamans and Kathy Thurston. Special thanks also to Lisa Yane for coordinating the scheduling, travel, and housing arrangements for the feline subjects and for obtaining medical clearances in connection with anesthetizing the research assistant.

REFERENCES

Boone P. Cat reactions to clean-shaven men. *West Musicol J* 1958; 11 (2): 4–21.

Ciccone M. L. Feline responses to hairy legs. *Midwest Sociolog Rev* 1986; 32 (1): 51–79.

Conroy G. Feline responses to ponytails. *Urban Sociol Rev* 1987; 21 (36): 302–321.

Norquist W. G. Feline reactions to Supreme Court nominees. *J Feline Foren Stud* 1988; 12 (8): 437–450.

O'Connor S., Brynner Y. Feline responses to shaven heads. *J Head Trauma* 1990; 42 (17): 309–324.

Quant M. Cat responses to bangs. *Tonsolog Proc* 1965; 3 (5): 251–262.

Schwarzenegger A. A study of how cats respond to body hair. *Mind/Body Rev* 1983; 3 (12): 25–108.

Seuss Dr. Feline responses to hats. *Vet Devel Stud* 1955; 32 (7): 54–62.

Smith J, Brothers J. Feline responses to healthy adults. *Health Advice* 1972; 51 (9): 32–33.

1. Bork was a nominee to the United States Supreme Court. Because of Bork's distinctive beard, his photograph has been used in reaction studies with dogs, rats, and planaria (flatworms), and in bombardment studies with pigeons.

2. We excluded all data pertaining to the 15 cats who gave birth while viewing the photographs. The reproducibility of the GAVE BIRTH/DID NOT GIVE BIRTH data will be addressed in a separate, future study.

3. Photographs do not, of course, produce pheromones, but they do emanate airborne scents derived from chemicals used in the developing process. Our study with animate bearded men will employ a strategy to eliminate this imbalance: Before being shown to the cats, each bearded man will be immersed in a photochemical bath.

THE BEARD-SECOND: NEW UNIT OF LENGTH

Kemp Bennett Kolb
The Haverford School, Haverford, Pennsylvania

Just as the astronomers have on a cosmic scale a unit of length related to the time something special travels at its own speed, physics has long awaited a corresponding unit in the microcosmos. The proposed unit is the BEARD-SECOND: the distance a standard beard grows in 1 second. Conveniently, there are nearly 1,024 beard seconds in one light year, placing the new unit in the virus particle range.

To complete the definition a standard beard is defined as growing on a standard face at a rate of 1 beard-second (100 angstroms exactly) per second.

An inside glimpse at what's new in emerging technologies

Stephen Drew

The Hair Recorder

It has long been known that a strand of hair can incorporate information about the environment in which it is growing. Researchers at Elbot Laboratory are developing a way to retrieve the audio information that sometimes becomes encoded in a strand of human hair. Potential uses include memory aides, historic research, and firsthand—or, more accurately, "firsthair"—evidence in legal proceedings.

Sound patterns can strongly influence the protein structure of the hair during stressful periods of hair growth (see Rebekah Sage's seminal article, "A Stress Analysis of a Bouffant," *JIR* 32:4). The "recording" is actually a chemical process that occurs in the hair follicle. An individual hair can be used to "replay" many of the sounds that impinged upon the follicle while the hair was growing.

Human hair has been shown to record better in young people than in their elders. Youngsters' thicker hair (often 80 microns or more in diameter, compared to a typical diameter of 60 microns in older people) has the capacity to encode copious amounts of information. This abundance of data, though, presents numerous difficulties in retrieving and decoding the sound information.

Most shampoos, pomades, and unguents degrade the quality of the recording. A few—especially shampoos containing fish oil or kerosene—seem to enhance hair's ability to capture and store detailed sound records.

The technology has "yielded some impressive results," according to Noah Morgan, Elbot Laboratory's Director of Research & Development. Morgan acknowledged that political pressures are affecting the laboratory's Aural History Research Project. The project involves hair samples from Abraham Lincoln, Confucius, Isaac Newton, Brigham Young, Aristotle, Moses, Cleopatra, Genghis Khan, Enrico Caruso, Napoleon, John Keats, Lady Godiva, L. Ron Hubbard, and other important personages.

Many of the objections come from religious authorities. "There is concern," says Morgan, "about the hair recorder being used to play back sounds from the tonsorial remains of early religious figures. This conceivably could reveal truths that might not square with the expectations of modern believers. It could be, pardon the expression, a hair-raising experience."

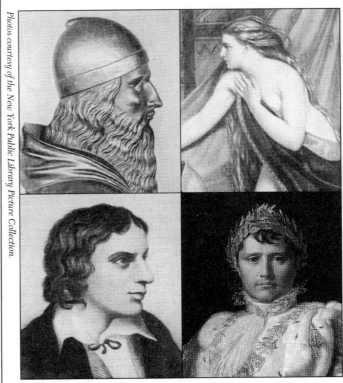

Photos courtesy of the New York Public Library Picture Collection.

Clockwise from top left: *Aristotle, Lady Godiva, Napoleon, John Keats.*

BALDNESS AND HYDROFRICTION

Joel I. Shenker and Neal Stolar

Department of Psychology, Beckman Institute, and College of Medicine, University of Illinois at Urbana-Champaign, Illinois

Introduction

We contend that baldness is primarily a product of behavioral, not genetic, factors. Specifically, we hypothesize that baldness occurs by chronic exposure to water impacting upon the scalp, most especially when showering.

Elaboration and Support of Theory

We contend that the primary behavioral mechanism responsible for baldness is exposing oneself to the persistent frictional stress created by the impact of water on the scalp. In particular, we suggest that the prototypic forum for chronic hydrofriction stress (CHS) is the common overhead bathroom shower. By extension, then, the CHS-balding hypothesis predicts that people who persistently face the front of the shower (or front-facing showerers) will go bald by developing a receding hairline that gradually ebbs back; those who repeatedly shower with their back to the shower nozzle (or back-facing showerers) will go bald by developing a bald spot in the back of the head that eventually grows and spreads frontward. Indeed, a survey of anecdotal reports (Kerr, 1977) confirms this prediction.

Biblical Balding

Interestingly, biblical scholars may have already suggested the CHS balding mechanism, albeit unwittingly. Biblical figures are almost always painted and otherwise portrayed with a full head of hair (eg, Charlton Heston as Moses). Since overhead showers are not mentioned in the Bible, it is reasonable to assume that such devices did not exist at contemporaneous times. Scholars have subconsciously recognized the absence of overhead showers in the Bible and naturally inferred an absence of tell-tale baldness in biblical characters.

Cross-Cultural Balding

Evidence supports a CHS-balding etiology in different cultures. For example, cross-cultural observations reveal that there are fewer bald people in Japan than in the United States. Apparently, the civic-minded presence of public baths in Japan (contrasted with a comparative paucity of same in the United States) has provided an alternative to the harsh hydrofrictional hair stress of the overhead shower, thus reducing baldness in the Japanese population.

Balding and the Sexes

Women lose their hair less often, more slowly, and less noticeably than men. These observations are explained by noting that, historically, women have been more likely than men to wear shower caps when showering, women have been more likely to take baths, and women have been more attentive to basic hair care (Druff, 1979).

Ontogeny of Balding

Observations of human development further support the CHS balding hypothesis. Neonates, for example, are occasionally seen with some hair on the scalp, which is quickly lost over the first few days of life (Shenker, 1990). Parsimony compels us to conclude that the viscous drag created by the outrush of amniotic fluid during delivery causes neonatal hair loss. We speculate that the neonate's hair is not fully habituated to the fluid-frictional stresses of the *ex utero* world (having previously experienced only the less turbulent *in utero* fluid flow). Thus, the hydrofrictional hair-loss threshold may be lower in the neonate, potentiating the acute balding effect.

Yet as infants develop, their young hair follicles recover from their hypertender neonatal state. As other new follicles emerge, youngsters quickly sport a full head of hair. This developmental step leads to still another relevant observation: After infancy, children are never (or rarely) bald. This fact is quickly explained by noting that children more often take baths than showers, thus minimizing the stress on their hair.

Typically, the earliest signs of balding do not begin until after adolescence, a developmental period that is accompanied by increased attentiveness to personal hygiene and hair care. The resulting sharp increase in the number of showers and the transition to harsher shampoos likely combine to damage hair that has been accustomed to the gentler and less frequent hydrofrictional stresses of the preadolescent childhood years. In addition, the inexperienced and often clumsy application of hair creams and conditioners with which people of this age often experiment creates further hair stress. The cumulative weight of all these factors must provide an overwhelming CHS effect, thus initiating hair loss. Further, the fact that hair loss does not immediately follow the postadolescent increase in hydrofric-tional hair torture is testimony to our hypothesis that *chronic* hydrofrictional stress is critical in producing baldness.

Conclusion

We have shown that baldness is most likely caused by unenlightened showering habits and is therefore unrelated to genetic mechanisms.

Acknowledgments

Several people made useful comments on earlier versions of this paper, contingent upon our assurance that they would remain anonymous. Their names are Marvin Alkin, Brenda Anderson, Cheryl and Curt Condon, Kathy Dwyer, Diane Lynn, Graeme McGrufficke, Ed Roy, Arden and Lois Shenker, Debra L.W. Shenker, and Pete and Naomi Watson.

REFERENCES

Druff Dan. Hair care habits. *Chron Hairier Educ* 1979; 77:433–444.

Kerr Fae. Bathroom exit polls. *Trends in Porcelain* 1977; 44:123–141.

Shenker Debra LW. Personal communication to JIS, 1990.

TECHNOLOGY UPDATE

An inside glimpse at what's new in emerging technologies

Stephen Drew

The NMR Toupee Detector

The toupee detector demonstrated recently by the National Research Bureau will likely be introduced to the commercial marketplace by midyear.

The technology is based on nuclear magnetic resonance (NMR)/wax paper engineering studies carried out at Stanford University and the National Institutes of Health in the mid-1980s.

Major commercial questions have centered on the human interface, specifically on the type of indicator to be chosen for the commercial form of the device. Preliminary research (sample size: 5) indicates that consumers prefer a form of indicator that emits high-intensity raspberry sounds or sirenlike warnings when in the presence of a cheap hairpiece.

The NMR toupee detector is currently undergoing Federally mandated shock testing.

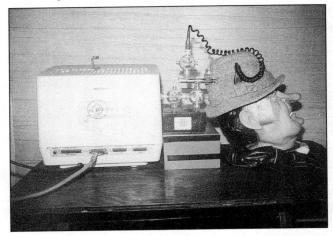

Contains 100% gossip from concentrate

Compiled by Stephen Drew

Hair Growth Breakthrough

A new hair growth medicine is showing great promise. Minixidahl, a topical preparation that is derived from orange juice, has caused lush, copious hair growth on more than 90% of human subjects. While the treatment is remarkably effective, it has one major drawback: On most test subjects the hair growth is limited to the outside of the nose (Figure 1). Some subjects also showed increased growth of internal nose hair.

Hot Side Hot, Cold Side Cold

A new power plant will marry the benefits of nuclear fission and fusion. In a novel design twist, the structure was inspired by a hamburger container. The facility is known as a "hot side hot/cold side cold" (HSH/CSC) power plant. It exploits a vortex interaction that develops when nuclear fission occurs in close proximity to nuclear cold fusion. The HSH/CSC power station is being built outside Salt Lake City, Utah. The containment facility was modeled after a McDonald's hamburger package. The original fast-food packaging was intended to regulate the temperature levels of a hamburger (the hot side) and a topping (the cold side) composed of lettuce, tomato, and condiments. A McDonald's ad of the 1980s explained that this kept the "hot side hot" and the "cold side cold." The technology underlying the design was inspired by the cold fusion research performed by Stanley Pons and Martin Fleischman at the University of Utah.

Nuke vs. Nuke

There is a new battleground in the intense competition for energy research funding. Proponents of the nucular energy camp are breaking away from their former colleagues, the nuclear energy researchers. The nucular/nuclear split is the latest sign of disharmony in a field that was once very tightly bound. Now the nucular groups will be competing against the research groups devoted to nuclear fission, nuclear cold fusion, and traditional nuclear hot fusion. Observers wryly point out a parallel to this kaleidoscopically complicating realignment in the realm of research politics: the changing views over the last 30 years about the existence and classification of subatomic particles in nature.

Of One Mind

The mind/body problem has long perplexed philosophers. It has now been solved.

Figure 1

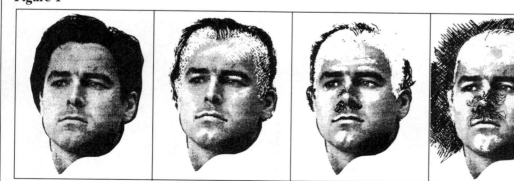

| DAY 1 | DAY 5 | DAY 10 | DAY 15 | DAY 20 |

Sidewalk Strut

Why do old ladies always walk in the middle of the sidewalk? That question has fascinated Patricia O'Leary-Savage since she was a 26-year-old graduate student. O'Leary-Savage, who is now chairman of the psychology department at Kansas Normal University, has been doing research on the question since 1964. She has reached some tentative conclusions.

Exciting Phone Calls

Many people are unaware that the word "laser" was originally an acronym for "Light Amplification by Stimulated Emission of Radiation." Now science and business are teaming up to take advantage of the stimulation that occurs when lasers are used to transmit telephone calls. Certain light frequencies have long been known to be stimulating or exciting when viewed by human beings. The psychedelic poster painters of the 1960s made great use of these colors. Researchers at MCI/Bell/KDD/British Telecom Consolidated Laboratories recently discovered that these light frequencies can also impart excitement to any signal that will later be converted to audio form. When the new erbium laser optical transoceanic communications cables are laid across the Atlantic (1993) and the Pacific (1996), telephone users should notice an added dash of excitement every time they make an overseas call.

Car Talk

It will soon be possible for a motorist to telephone the driver of another automobile by dialing that car's license plate number. The new "super-cellular" technology will be available in California later this year.

All Bent Out of Shape

In 1928 an Englishman named Frank Plumpton Ramsey proved mathematically a surprising fact that, 63 years later, is causing fistfights in university taverns throughout the world. The controversy pits theologians against astronomers. Ramsey showed that if you have a very large number of randomly distributed stars (or pebbles, numbers, or whatever), you can expect to find, somewhere in their midst, almost any particular pattern you can imagine. It was discovered in recent years that the universe contains large empty regions and gigantic wall-like patterns of stars. Some (though not all) astronomers feel this is explained simply by Ramsey's theory. Some (but not all) theologians feel it is evidence of God's design.

Pickled Socks

A new study indicates that pickled socks may not be a cancer-causing agent after all. The study, conducted with more than 3,500 male subjects over a period of 12 years, appears to contradict earlier findings.

SLEEP RESEARCH UPDATE

- SD is sleeping with DE, GS, JJ, WL, EL, PdeR, GO, CS, KU, TN, RR, RL, WJ, and SP.
- GW is sleeping with HK and an air conditioner.
- JV is sleeping with WB.
- NI has stopped sleeping with LH for logistical reasons.

Public Believes Scientists

The public usually believes anything—no matter how foolish—that a scientist says, according to recent research performed by Les Marsden of the J.T. Spaulding Institute. Marsden conducted a series of tests with members of the general population. The subjects believed they were answering an opinion survey about "recent scientific discoveries." Actually, they were responding to blatantly nonsensical "facts" that Marsden had concocted.

Among Marsden's findings: 78% of the subjects believed that Venus orbits around Jupiter after they were told that there is scientific proof. Before being told about the supposed scientific proof, only 42% of subjects believed this. After they were told that there is scientific proof, 84% of the subjects believed that reading books causes cancer. Before being told about the scientific proof, only 55% of subjects believed this. After hearing that there was scientific proof, 63% of subjects believed that apes are descended from trees. Before being told about the "scientific proof," only 39% of subjects believed this.

CHAPTER 13

SNEAKING UP ON CANCER

THE TRUE FACE OF A CANCER CELL

Johan Moan
Institute for Cancer Research
Oslo, Norway

Among scientists engaged in cancer research it is well known that the true nature of a cancer cell is rarely revealed. Cancer cells are adept at hiding their faces from investigators. It is almost impossible to get them out of their hiding places among normal cells by means of monoclonal antibodies or other sophisticated bait. In all types of microscopes they appear very similar to normal cells. Our delight was therefore enormous when we were able to take a photomicrograph of cancer cells at a moment when they revealed their faces. Such a picture is here published for the first time.

Materials and Methods

A Leitz Diavert microscope with a phase-contrast objective was used. A brief flash of light was given to expose the Kodak film. The NHIK 3025 cells are derived from a cervix carcinoma *in situ* and cultured in centaur serum (20% horse + 10% human) as earlier described (Oftebro and Nordbye, 1969). Sometimes, when grown in this way on bottles sterilized with ethanol, these cells can be induced to show their true nature.

Results and Discussion

Figure 1 shows a face contrast picture of true cancer cells. It can be seen that two cancer cells share one face. Furthermore, the nose and the tip of the tongue are located on one cell, while the brain is on the other cell. (It is not an uncommon biologic phenomenon that good brains and good tongues are located on different individuals.) To help the reader visualize the cells, we have equipped them with an artificial hat and an Italian bow tie. This does not coarsen the results, since all hats are artificial and most bow ties are Italian.

Noted should be the wicked expression in the eyes and a large number of pigmented nevi on the nose and on the cheek. This is in agreement with earlier published papers, which shows that cancer is malignant (Aristotle, 329 BC) and that certain tumors (e.g., melanomas) frequently arise from pigmented nevi (Conti et al, 1989).

Acknowledgment

The present work was supported by the Norwegian Association for Horseracing.

Figure 1

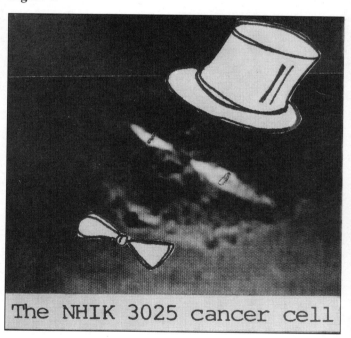

The NHIK 3025 cancer cell

REFERENCES

Aristotle. *Organon,* 329 B.C.

Conti CJ, Slaga TJ, Klein-Szanto AJP, eds. *Skin Tumors: Experimental and Clinical Aspects.* New York: Raven Press, 1989.

Oftebro R, Nordbye L. Establishment of four new cell strains from human uterine cervix. *Exp Cell Res* 1969; 58:459.

HOW TO FIND THE CURE FOR CANCER

Edmund A. Gehan
Houston, Texas

The cure of cancer has been sought for centuries. Since the late 1930s, this search has been coordinated in the United States by the National Cancer Institute (NCI). No cure for cancer has yet been found, so I venture to propose a cure for cancer and a method for its evaluation. My proposal will put some people out of work because more people are living off the search for a cancer cure than dying from it. But so be it; too much money has already been spent without success. Because I am a statistician, my approach necessarily has statistical elements: postulates, axioms, and other aids to reflective thinking.

My first postulate is from Leonardo da Vinci: "No human investigation can be called real science if it cannot be demonstrated mathematically." Leonardo da Vinci (1452–1519)

My second postulate is from another eminent authority: "Don't bet on finding a cure for cancer unless you know the odds." Jimmy the Greek (b. 1921)

Proceeding directly from these postulates, I now state my first axiom:

AXIOM I: *Cancer will be cured by the best treatment.* You might say that the axiom is obvious, perhaps even a self-evident truth, but then isn't that the definition of an axiom? I begin by setting down a list of currently available treatments: surgery, radiotherapy, chemotherapy, immunotherapy, and "other" treatments (e.g., vitamins, Laetrile, Krebiozen, and prayer). The "other" category is a catchall for possible cancer cures. After all, NCI has used biochemical formulations, random searches, synthetic analogs, and thousands of other treatments without finding the cure, so I here consider the possibility that the cure is "other" treatment.

To evaluate the treatments, I consider a 4^5 factorial design of study (Table 1). There are five factors corresponding to the treatments, so the study will be multimodal, designed to appeal to all and offend none. Next, I consider how much of each treatment: none, a little, some, and a lot. Not very precise, you might say. Well, my reply is, "So what?" With five factors, each at four levels, it is simple to work out that we have a 4^5 factorial experiment with 1,024 treatment combinations. One example is surgery (a little), radiotherapy (a lot), chemotherapy (some), immunotherapy (none), and Laetrile (a lot). There are 1,023 other possibilities. It is time to apply:

AXIOM II: *The best protocols have the best treatments in them.*
With 1,024 treatments we surely have at least *one* good

Table 1. Factorial Design of Clinical Study.

Factors (5): Surgery, radiotherapy, chemotherapy, immunotherapy, other
Levels (4): None, a little, some, a lot
Experiment: 4^5 factorial, 1024 treatment combinations
Example of treatment: Surgery (a little), radiotherapy (a lot), chemotherapy (some), immunotherapy (none), Laetrile (a lot)

Table 2. Protocol.

Objective: To cure cancer
Selection of patients: Patients will be chosen at random from the population of all patients with cancer
Design of study: 4^5 factorial
Treatments: There are 1,024 treatment combinations
Pathology: All patients must have pathologic confirmation of diagnosis
Endpoint: Disappearance of all tumors for at least 5 years
Statistical considerations: n patients on each treatment to ensure given significance level and power of test

one; the problem is to find and evaluate it. The next step is obvious: Write a protocol. Table 2 gives the key elements. Since we have a complex factorial design, over 1,000 treatment combinations, patients selected from all over the place, pathologic confirmation, a stringent end point, and *n* patients needed on each treatment, I now call on Axiom III:

AXIOM III: Don't give up when the going looks tough. When using statistical expertise, the solution is to fractionate the factorial design. Since I have a 4^5 factorial, I fractionate symmetrically by taking a $1/5^4$ replication of the 1,024 treatments. By the beauty of statistical theory this leads to two treatments—a nice convenient number. A remaining problem is to decide the number of patients to study on each. I have heard that 14 patients[1] should be studied to evaluate each new treatment; however, on this protocol we would like to be *twice as sure* to find the cancer cure, so we will include 28 patients. This is not a bad number or age, and, in fact, I saw this number come up three times in the lottery last week. This brings me to my third postulate: Not only is it important to be good, but it helps to be lucky, too!

I now decide to enter the 56 patients into a randomized clinical trial comparing CRISO treatment (Chemotherapy, Radiotherapy, Immunotherapy, Surgery, and Other) vs. placebo. It will be *prospective, randomized, quadruple-blind,* and *unbiased.*

The NCI could not possibly disapprove of a study with so many randomized elements, so let us look into the future and assume the results in Table 3. Note that the clinical trial has exactly 28 patients on each treat-

Table 3. Results of Clinical Trial.

	CRISO	**Placebo**
Registered	28	28
Eligible	14	28
Evaluated	14	28
Too early	13	0
Fully evaluable	0	14
Partially evaluable	1	14
Not evaluable	0	0
Responders	1	1
Response rate	1/1 (100%)	1/28 (4%)

ment, which is already a landmark; it is the first clinical trial to have precisely the number of patients specified in the protocol.

We also see that the number of eligible patients on CRISO is only 14 compared to 28 on placebo; 13 have been considered "too early" on CRISO, while none were on placebo. There has been one responder on each treatment. Because there is only one partially evaluable patient on CRISO and 28 evaluable patients on placebo, the response rate is 100% of CRISO and 4% on placebo! A careful statistical evaluation using Fisher's exact test gives the significance level $P = .07$. The crucial step in achieving 100% response on CRISO and 4% on placebo was:

AXIOM IV: If at first you don't succeed
     And if the response rate
     Is not what you need
     Evaluate! Evaluate!

Evaluate is what I have done—by noting that 14 patients on CRISO had advanced disease and were therefore not eligible for CRISO. After all, we must start curing cancer in those with limited disease. Next, though 13 patients received CRISO for six months with no effect, I classified them as "too early" because there is still hope for response. Hence I now have only one patient on CRISO, and she thought it was very beneficial. There were no problems for the 28 patients on placebo, and with only one responder the response rate was 4%. Unfortunately, this result was still not statistically significant at the .05 level. Something else was needed, so I decided to use:

AXIOM V: Statistical significance is everything.
     Unless P equals .05 or less
     Who could care less?
     When P is not the proper size,
     Analyze! Analyze!

I began to look for subgroups of patients in whom the response to CRISO is especially beneficial and, lo and behold, obtained the data in Table 4. This study was conducted in Las Vegas, and the patients could be divided into two groups: blue-eyed brunette Las Vegas showgirls and the "other group." The responding patient on CRISO was a blue-eyed brunette Las Vegas showgirl. Placebo, unfortunately, was of no help, and the response rate was 0% in 27 patients. Now, a careful application of Fisher's exact test yielded a significance level of

Table 4. Response Rate by Patient Characteristic.

Patient Characteristic	CRISO	Placebo	P
Blue-eyed brunette Las Vegas show-girls	1/1 (100%)	0/27 (0%)	.04

$P = .04$. The magic barrier of $P = .05$ had been broken, and the cure of cancer was a real possibility.

Of course, the ultimate criterion for cure of cancer is living for a long time, so Figure 1 gives survival curves by response status for CRISO and placebo. The one patient who responded to CRISO has lived for 5 years,

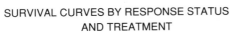

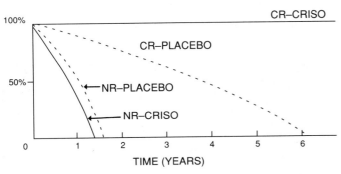

and so there is 100% survival. Nonresponders on both treatments have not done very well, and these patients died in a median time under one year. So survival on CRISO is clearly better than that on placebo!

Figure 2 shows that this search for the cancer cure has followed the precepts of the scientific method. We

Figure 2

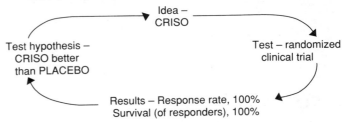

began with the combined multimodality treatment CRISO, there was a test by a randomized clinical trial, the results of response rate and survival are both 100% and our statistical test demonstrated that CRISO was significantly better than placebo.

Our methods have been completely scientific and quantitative, so we summarize by:

AXIOM VI: The cure for cancer can be found
By methods scientific
By treatment specific
By doctors terrific
In patients characteristic
Analyses hieroglyphic
Conclusions that stick
In short, by STATISTICS!

In fact, through a completely scientific, objective, and analytic methodology, it has been established that: *Statistics is the cure for cancer (Q.E.D.).*

REFERENCES

1. Gehan EA. The determination of the number of patients required in a preliminary and follow-up trial of a new chemotherapeutic agent. *J Chron Dis* 1961; 13:346–353.

SUNLIGHT CAN PREVENT SKIN CANCER

Johan Moan

Institute for Cancer Research, Oslo, Norway

In the following, it will be proven that the scientific proof for the theory that sunshine is dangerous is based solely on lack of knowledge of the sine relationship among epidemiologists.

Materials and Methods

The present work is built on data for skin cancer incidence rates at different latitudes in Norway provided by the Norwegian Cancer Registry. Only data for women have been considered, since they are more than sufficient and since there is a law in Norway saying that whenever possible, women should have an advantage over men.

The sine relationship[1] was used.

Results and Discussion

The amount of sunshine falling on a female body at different latitudes, L is shown in Figure 1. We have chosen to show the situation at the vernal (or autumnal) equinox, since it represents an average of the year. It should be noted that in the daytime, most women in Norway stand upright with their axis in vertical position.[2] Thus the sun exposure is simply $S \sin L$ where S is the solar constant and L is the latitude.

Figure 2 shows the incidence rate R of squamous cell carcinomas of women at different latitudes L in Norway as a function of sun exposure $S \sin L$. It can be seen that there is a strict linear relationship between the incidence rate R and $\sin L$. The superiority of the theory is demonstrated by the fact that the regression line passes through the point $\sin L = 1$, $R = 0$. This is in perfect agreement with the observations: Sin L is the North Pole, and there has never been observed a single case of skin cancer in women at the North Pole. Thus, thanks to the sine relationship, we have proven that sunlight protects against skin cancer.

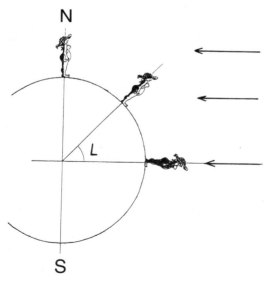

Figure 1. The situation among women on the earth at vernal equinox. During daytime, even at vernal equinox, most women in Norway are oriented with their axes vertically. L is the latitude.

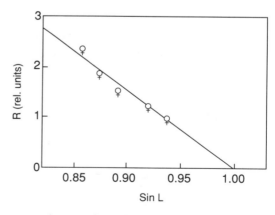

Figure 2. The age-adjusted incidence rate R of squamous cell carcinomas in women in Norway as a function of $\sin L$, where L is the latitude at which most women live.

1. Probably dating back to Pythagoras.
2. Data from the Norwegian Statistical Association.

HOLISTIC DIAGNOSIS: RABIES AND CANCER

Ulla Iversen and Olav Hilmar Iversen
*Institute of Pathology, University of Oslo,
Oslo, Norway*

There is general agreement that cancer can be caused by chemical, physical, or biologic carcinogens. The biologic carcinogens are viruses. The problem of determining the etiology of a particular case of human cancer is much more difficult and can only rarely be solved with certainty.

Holism has become a highly influential doctrine in modern medical science. Holistic theory is based on the self-evident premise that all details must be understood in the context of the whole. Thus all symptoms and signs in the *whole* body must be examined before a diagnosis is attempted; otherwise, the diagnosis will be incomplete. We have it on the highest authority (Boss, 1978) that the body is not an object or a thing, but a state or a condition. The physician who learns to interpret the body's silent language will be in the best position to understand the disease. Thus a sarcoma, for example, may be a symptom of a more profound malfunction, such as depression, leading to reduced immunological resistance.

Case Report

A 40-year-old female was presented with a retroperitoneal mass. Histology showed a sarcoma with cellular and nuclear atypia and several atypical mitoses. One of these confirmed direct anecdotal evidence of the etiology. About 35 years previously (Figure 1), the patient had lived in an area where rabies was prevalent and had been bitten by an infected poodle. The dog was killed, and the patient was given anti-rabies treatment. Yet 35 years later, she developed a tumor. The morphologic evidence from the mitosis, as shown in Figure 2, confirms the patient's own experience.

Discussion

According to holistic theory, anecdotal evidence is evidence nonetheless. Hence we need to establish some link between the dog and the sarcoma, since there can be no doubt that the nuclear morphology points directly to the poodle. It seems highly probable that this sarcoma is viral, induced by a rabies virus transmitted to the patient through the dog, whose ferocious attack and

Figure 1. Snapshot from the family album. The little girl and her mother on their way to meet the father. The girl's interest in the poodle is obvious.

subsequent killing can have precipitated an emotional trauma that had an adverse effect on the patient's immune status. This is supported by the fact that the patient was not given counseling. Thus the purely somatic treatment aimed only at a single symptom, the rabies, achieved only a partial cure. The fact that the patient's whole life was not examined meant that the disease was probably forced underground, to reappear years later in the form of an oncogenic virus. Fortunately for medical science, the histologic section from the tumor not only told us that the patient suffered from a sarcoma, but also provided clear evidence of the canine etiology.

It is well known that some viruses are oncogenic. Rabies virus is an RNA virus. A well-known morphologic sign of virus infection is inclusion bodies in the nuclei, but here we propose that the form of the mitotic figures may also be significant. This presentation also constitutes exciting evidence that holistic medicine can be applied on many levels, including the morphologic one. This may well be the first step toward a new branch, holistic morphology, based on pattern recognition. The perspectives opened up by such a discovery are endless, not to say mind-boggling.

A further, more paramedical, perspective opened up by this finding is the light that it sheds on the tricky question of the extent to which Nature imitates Art, or vice versa. This is supported, or vice versa, by McWhinney (1978) in an excellent article in which he says:

"The scientist's symbols are technical terms . . . which bear a *direct relationship*° to the objects or facts which they symbolize. The artist's symbols are metaphorical forms, verbal and nonverbal, which express their meaning *indirectly and obliquely.*"[1]

Here we have an example of a metaphoric nonverbal form (the chromosomal morphology) expressing a scientific fact (the poodle). We can thus infer that the relationship is both direct and oblique.

Conclusion

If we train ourselves to be aware of every possible expression of body language, *including* that expressed in histologic sections, and if we are at the same time artistically inclined, we may discover the etiology of some hitherto unexplained diseases. This case has provided morphological evidence of the cause of a malignant disease. It is the most direct example of body language that the authors have ever encountered in their careers as histopathologists.

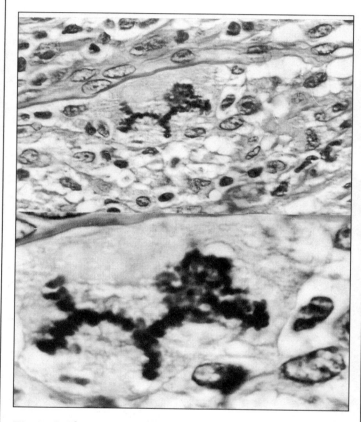

Figure 2. Photomicrograph of a mitosis in a sarcoma showing how the organization of the chromatin reflects the cause of the disease. The upper panel shows a picture with initial magnification objective 40, and the lower panel shows the same with objective 100. Both images are from the same poodle.

REFERENCES

Boss M. *Praxis der Psykosomatik: Krankheit und Lebensschicksal.* Bern: Benteli, 1978.

McWhinney IR. Medical knowledge and the rise of technology. *J Med Philos* 1978; 3:293–304.

1. With apologies to McWhinney for taking his words entirely out of context, and for inserting our italics°.

CHAPTER 14

MEDICAL WONDERS

INFECTIOUS DISEASES IN BRICKS

Ruthifer Karron, M.D.

The Thayer Memorial Clinic, Salt Lake City, Utah

The brick is a clay-based rectangular organism. It is found chiefly in and around human settlements. We conducted an extensive survey of infectious diseases among the brick population in and around Salt Lake City. The population showed an unexpectedly high incidence of syphilis.

The Survey Population

The total estimated brick population of the region is 650,000,000,000. We conducted the survey among a sample subpopulation of 120,000,000 bricks.

Incidences of Infectious Diseases

Infections of the central nervous system:

Meningitis: 0   Encephalitis: 0

Infections involving the cardiovascular system:

Endocarditis: 0   Pericarditis: 0
Myocarditis: 0   Rheumatic fever: 0

Infections of the ear, nose, and throat:

Otitis media: 0   Ludwig's angina: 0
Malignant external otitis: 0   Pharyngitis: 0
Mastoiditis: 0   Epiglottitis: 0
Sinusitis: 0   Herpangina: 0
Parapharyngeal and   Vincent's syndrome: 0
    retropharyngeal   Oral candidiasis: 0
    infections: 0

Infections of the lower respiratory tract:

Bronchitis: 0   Pyogenic lung abscess: 0
Pneumonia: 0   Empyema: 0

Intra-abdominal infections:

Tuberculous peritonitis: 0   Pancreatitis: 0
Diverticulitis: 0   Enteric fever syndrome: 0
Cholecystitis: 0   Hepatitis: 0

Sexually transmitted diseases:

Gonorrhea: 0   Chlamydia: 0
Syphilis: 1

Other:

Smallpox: 0   Cholera: 0
Measles: 0   Rubella: 0
Mumps: 0   Rabies: 0
Influenza: 0   Tetanus: 0
Diphtheria: 0   Plague: 0

Summary

Bricks may be prone to contract syphilis.

REFERENCES

American Academy of Pediatrics. *Report of the Committee on Infectious Diseases*, 19th ed. Evanston, IL: Academy of Pediatrics, 1982.

Benenson AS. *Control of Communicable Diseases in Man*, 12th ed. New York: American Public Health Association, 1980.

THE EFFECT OF CENTRIFUGATION ON HIP REPLACEMENT PATIENTS

E.L. Radin, M.D.,[a] A.L. Shiller, M.D.,[b] R.M. Rose, Sc.D.,[c] and A.S. Litsky, M.D.[d]

From the [a]University of West Virginia, Morgantown, West Virginia; [b]Massachusetts General Hospital, Boston, Massachusetts; [c]MIT, Cambridge, Massachusetts; [d]Mt. Sinai Hospital, New York, New York.

Centrifugation of PMMA bone cement increases its fatigue life (Burke et al., 1984). Pressurization of the cement retards the loosening process and thus improves the prosthetic fixation (Oh et al., 1978). In exploring the hypothesis that centrifugation of the cement during polymerization would confer further increases in fatigue resistance, we have discovered a procedure that simultaneously augments the insertion pressure of the cement.

Materials and Methods

A standard AHS operating table (No. 007) was modified to spin about the vertical axis at the center of gravity. The tabletop was fitted with accelerometers at the extremities and at the center (recessed to avoid discomfort to the patient). A finite element analysis indicated the rigid fixation of the patient would be essential. This was achieved with nylon webbing over the thorax, pelvis, contralateral leg, both arms and shoulders, and head. Due to the rotation of table and patient, endotracheal anesthesia proved difficult; standard spinal anesthesia therefore proved to be the method of choice. (Our experiences with hyperbaric spinal anesthesia will be reported in a subsequent publication.) The intravenous lines were brought down over the center of rotation of the table, through special rotating couplings.

Patients were operated on in the decubitus position using standard total hip replacement techniques and the Harris Hip System, using low-viscosity PMMA cement and special self-locking jigs. Immediately following proper insertion of each prosthetic component, the patient was spun at 3,600 rpm (60 Hz) until the cement had hardened.

This procedure was carried out on 20 adult patients (19 women and 1 man) with an average age of 65.3 (range, 59–71). The male patient (the first subject) sustained fractures of the ipsilateral femoral shaft, contralateral proximal tibia, and surgical neck of the contralateral humerous and elbow, necessitating redesign of the restraint system. There were no injuries to the subsequent patients or operating room personnel. The remaining patients were followed for between 2 and 3½ years with an average of 2¾ years. Their stay in the hospital was apparently longer than the usual, but the increase was not statistically significant. All complained of some initial difficulties with balance, which cleared within six to nine months of the procedure.

Clinical Results

Results were evaluated at 3, 6, 9, and 12 months, and

Figure 1

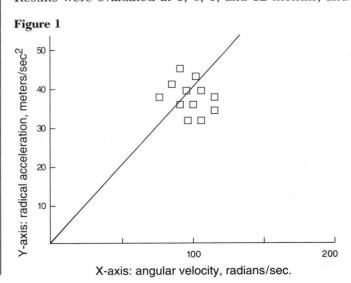

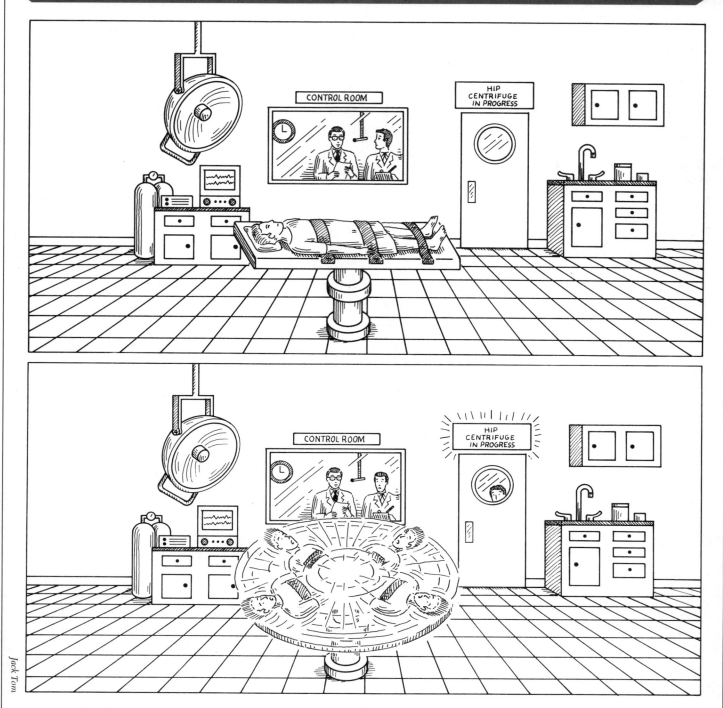

Jack Tom

semiannually thereafter. Radiographic analysis using a laser-follower photogrammetric system (Data Technology, Inc., Woburn, MA) revealed an absence of radiolucent lines surrounding the prosthetic components consistent with results of previous investigators (*loc cit*) who used passive pressurization methods. Our results also indicated a correlation between radial acceleration and angular velocity. Although there was considerable scatter, multiple regression analysis indicated that straight line was best, as shown in Figure 1. These re-

sults suggest a direct relationship between radial acceleration and tangential velocity for a rotating body.

We conclude that patient centrifugation can improve the fixation of total joint prostheses. The long-term effects have not yet been evaluated.

REFERENCES

Burke DW, Gates EI, Harris WH. *J Bone Joint Surg* 1984; 66A:1265–1273.
Oh I, Carlson IE, Tanford WW, Harris WH. *J Bone Joint Surg* 1978; 60A:608–613.

CAKUS CHOCOLATUS AND THE TREATMENT OF DISEASE

Jay M. Pasachoff, Naomi Pasachoff, and Eloise H. Pasachoff

Bronfman Science Center, Williams College, Williamstown, Massachusetts

It is widely rumored that a large number of common diseases and ills respond favorably to a treatment of chocolate cake, *cakus chocolatus*,[1] taken internally once or twice a day, usually after meals. In order to test this hypothesis, a series of experiments were set up in our laboratory (converted from our dining room for the purpose).

Subjects were recruited randomly from the local population of Williams College students. Every third passerby was approached with the question: "Would you be willing to participate in a laboratory experiment to test the efficacy of chocolate cake in curing whatever ails you?" It was found that 97% of those asked responded affirmatively to both this question and to the follow-up question, "Do you have some minor ill that would make you eligible to participate in this experiment?" The other 3% all seemed torn, and paused before answering, but eventually muttered the word "diet," spun around, and ran off.

Subjects were divided into three groups. The first group consisted of six students and the first two experimenters cited above (who, as devoted to this study as any scientists, were willing to use themselves as guinea pigs in the name of science). They were fed cubic pieces of chocolate cake, *cakus chocolatus*, and were asked to consume it in a 3-minute period, along with their choice of coffee, tea, or milk. Each of the subjects was also supplied with a spoonful of sugar (to help the medicine go down).

A second group of equal size, the control group, was seated at the table, and then given empty plates, forks, beverages, and spoonfuls of sugar. As it was a single-blind experiment, they were not told whether they were the prime group or the control group, although some may have guessed from the difference in taste.

After a suitable wait, to allow the *cakus chocolatus* to act, all the subjects were asked how it had affected them. All the group one subjects felt that their general well-being had been aided by the ingestion of chocolate cake, and all group two subjects felt decidedly worse off than they had been a short time before. We interpret this to mean that, on at least a short-term time-scale, *cakus chocolatus* has a beneficial effect.

A third group was used to test topical applications of the *cakus* remedy. Three subgroups of group three were formed, totaling the size of the primary group. The chocolate cake was applied to the first subgroup in an axillary manner, with the measured cube of cake held under the left armpit for 60 seconds. None of these subjects reported the same beneficial effects that those who had taken the *cakus* orally had reported.

Another subgroup had the *cakus* applied facially, in the manner illustrated so well in the work of Keaton (Buster Keaton, films, 1917 et seq). They reported that overall they preferred Boston cream pie.

Subjects in still another subgroup, who had the *cakus chocolatus* applied anally, refused to answer any questions from the experimenters, and in fact have not spoken to us since.

The most junior author of this article (E.H.P.) was 16 days of age at the time of these experiments, and received her dose of *cakus* via her mother, the second author, a lactating female (N.P.). Her sucking reflex seemed to be stimulated.

As a result of these experiments, we can unhesitatingly recommend the oral application of *cakus chocolatus* as a general remedy, beneficial to all ages.

---

1. Although the generic name is given in this article, we actually used *cakus chocolatus* from Sara Lee Laboratories, which nobody didn't like. The relative efficacy of various competing brands of this powerful medicine is the subject of a current series of experiments that we are carrying out. As we are not restricting our experiments to the possible effects of these remedies on nonmalignant disease, we shall apply to the National Cancer Institute for grant support.

ODD REQUESTS FOR RADIOGRAPHY

Tim Healey, M.D.
South Yorkshire, England

The following comprised the "clinical information" justifying (?) the radiography of the body parts indicated. They originated in Saudi Arabia (whence they were transmitted to me by a radiographer who wishes to remain anonymous) and are sometimes amusing despite the language barrier. I wish to emphasize that my knowledge of Arabic is minimal, so I am in no position to laugh *at* those errors of syntax and vocabulary but hope that the originators have a sufficient sense of humor to allow me to smile *with* them. In each example the examination requested is bracketed.

1. "Captain in the Saudi Air Force; sometimes passes out during flight" (computed tomography [CT] scan of brain).
2. "Infertility" (chest; KUB = kidneys, ureters, bladder). The patient was 78 years old.
3. "Forgetfulness" (CT scan of brain).
4. "Headache for 30 years" (CT scan of brain).
5. "Obliques of T7-L3 to see little dogs and ducks." Oblique views of the *partes interarticulares* of the lumbar spine show the appearance of "Scottie dogs" (Figure 1).
6. "Patient trapped by head in automatic doors" (skull X-ray).
7. "Camel bone" (chest X-ray). Presumably it was swallowed.
8. "Swelling at the posterior aspect of the neck almost fixed to the skin with feeling of something eating her neck" (X-ray of cervical spine).
9. "Road traffic accident" (right and left thighs). The patient had had previous amputation of the left lower limb at the hip.
10. "Painful heels and decreasing erection" (lumbar spine and both heels). This makes sense. Gonorrhea and chlamydia can cause these symptoms and might show changes in the sacroiliac joints and in the heels.
11. "Patient breathed in flies. ? flies in sinuses" (CT scan of sinuses).
12. "Painless hematuria from the south" (excretion urography).
13. "Fell on her back when she was 12 years old. Now she has a backache" (lumbar spine X-ray). The patient was 60 years old.
14. "Patient was mixed in a concrete mixing machine" (magnetic resonance image [MRI] of brain).
15. "Hyperstimulation following personal therapy" (ul-

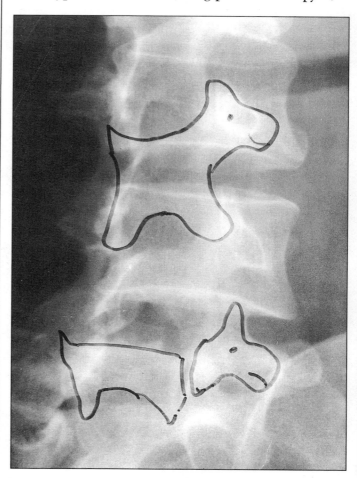

Figure 1. The 45° oblique view shows the normal shape of a "Scottie dog" (above); the dog has developed a collar where the bone is defective.

trasound examination of abdomen). To confirm pregnancy.

16. "Clinically nutcracker esophagus" (barium swallow for esophageal function). There is a condition known as "corkscrew esophagus," which is presumably what was meant.

17. "Work-up for a heart transplant" (ultrasound examination of right upper quadrant). The patient was 87 years old.

18. "History of spilt boiling water on chest" (chest X-ray).

19. "Revolving teeth" (ortho-pantomogram).

20. "? pituitary dwarf. Short for height" (MRI of pituitary fossa).

21. "Post-amputation" (X-ray of right hand).

22. "Whistling sound coming from both arms ? from elbows or wrists" (X-rays of both forearms).

23. "Patient bitten a wall" (skull X-ray).

24. "Camel kick to the frontal bone of the head" (X-ray of chest).

25. "Following amputation with a hammer" (X-ray of wrist).

26. "Constipation ? cause" (brain enema). Barium enema meant.

27. "Juice mixer explosion" (X-ray of right forearm).

28. "Backache after carrying a camel" (X-ray of lumbar spine).

NOBEL THOUGHTS

Profound insights of the laureates

Marc Abrahams

Sir John Kendrew is a member of the Medical Research Council Laboratory of Molecular Biology in Cambridge, England. In 1962 he was awarded the Nobel Prize in Chemistry for his work in determining the structure of the complex protein myoglobin. We spoke during a conference at Woods Hole, Massachusetts.

How often do you ride motorcycles?

The last time was during the war in what is now called Sri Lanka. I'd never ridden a motorcycle before. I was scared stiff. With a motorcar, if you take your hands off it, it keeps going, but with a motorcycle you can't be so sure. I found this so scary that I took it back to the transport people and said, "For God's sake, give me a Jeep." And they did. So how often is: once in a lifetime.

Do you own a leather jacket?

Yes, I do. In the English climate you need it from time to time.

If you had a tattoo, what would it be?

What I deplore about tattoos is that they're virtually irremoveable. What that suggests is that they're about something you're never going to change your mind about. And I can't think of anything like that.

Do you have any advice for young people who are entering the field?

I can tell you the advice I got from my advisor, [Lawrence] Bragg: For God's sake don't spend too much time reading the literature, because if you do, you'll find it's all been done before.

Of course, the other advice is get into a field that nobody else is in, because then you can work quietly and do whatever you want to do. And there's something to be said for working on a problem that seems to be impossible to solve, because then probably there won't be anyone else working on that problem, and there won't be people breathing down your neck all the time and racing for publication and all that stuff.

REDUCED SMULLYAN EYE CHART
Visual Cortex Impairment Test

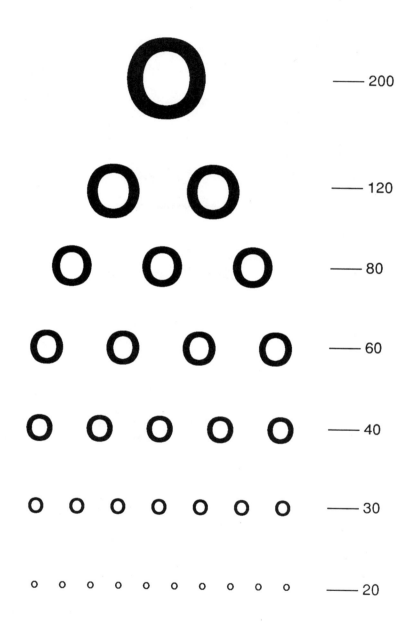

— 200

— 120

— 80

— 60

— 40

— 30

— 20

Visual Cortex Impairment Test

This chart allows an eyecare professional to assess whether a patient has suffered gross neurological damage. Patients with normal vision will recognize that all the letters are identical.

This chart was created by The Society for Basic Irreproducible Research. It is based on research performed by John J. Meagher, Jr., O.D., of the Harvard University Health Services Eye Clinic. Designation at side of line represents Visual Acuity expressed in Smullyan notation.

REDUCED SMULLYAN EYE CHART

Visual/Cognitive Proclivities Test

Q	— 200
C G	— 120
U V Y	— 80
B P F R	— 60
H N X Z N	— 40
O Q D G C Q G	— 30
Q G O C G C O D G C	— 20

Visual/Cognitive Proclivities Test

This chart allows an eyecare professional to assess a new patient's psychological proclivities, prejudices and biases. Patients with normal vision will recognize that, within any particular line, the letters can be easily mistaken for each other.

This chart was created by The Society for Basic Irreproducible Research. It is based on research performed by John J. Meagher, Jr., O.D., of the Harvard University Health Services Eye Clinic. Designation at side of line represents Visual Acuity expressed in Smullyan notation.

MORE ABOUT ZIPPERS . . .

X. Perry Mental
Ness Ziona, Israel

One of the first issues of *JIR* contained several articles concerning zippers (*JIR*, 1956; 3:3–9.), including one by Azo-Koh entitled "The Chemical and Biological Implications, Applications, and Complications of Zippery Mechanisms." In it, under the heading Zipper Applications in Medicine, the author stated:

> In surgery the plastic zipper will be a boon to forgetful surgeons. Instead of suturing incisions in the abdominal wall of operated patients by catgut, a plastic zipper mechanism will be installed.[1] Cases of clamps, retractors, scissors, tampons, etc., forgotten in the abdominal cavity will be promptly and easily dealt with by simply opening the zipper and by extraction of the forgotten paraphernalia. A word of caution is, however, necessary here: the end of the zipper should be provided with a locking mechanism, which would be unlocked only by the authorized personnel. This precaution is very important, inasmuch as the use of Freudian techniques for testing one's own interior machinery has already been attempted with disgusting results.[2]

Azo-Koh's irreproducible zipper predictions 30 years ago became a reality in 1986. B.J. Rubin of YKK in Macon, Georgia, writing in the **New England Journal of Medicine** of Nov. 6, 1986, p. 1234, warns and urges physicians not to use zippers for medical purposes because they are not sterile and may contain all sorts of oils, detergents, and other chemicals. Rubin ends by saying that he cannot be responsible (as a manufacturer of zippers) for any injury caused by use of zippers for medical purposes.

This letter is followed by a statement of Dr. Harlan Stone of the University of Maryland that the use of zippers "has greatly facilitated re-exploration of the

JIR RECOMMENDS
Articles, books, and other communications that merit your attention
Compiled by Stephen Drew, Norman D. Stevens, and X. Perry Mental

"Rectal foreign bodies: case reports and a comprehensive review of the world's literature," by D.B. Busch and J.R. Starling, *Surgery* 1986; 512–519. The comprehensive review is especially recommended. It cites reports of, among other items, seven light bulbs; a knife sharpener; two flashlights; a wire spring; a snuff box; an oil can with potato stopper; 11 different forms of fruit, vegetables, and other foodstuffs; a jeweler's saw; a frozen pig's tail; a tin cup; a beer glass; and a remarkable ensemble collection consisting of spectacles, a suitcase key, a tobacco pouch, and a magazine. Interested readers can request a reprint from James R. Starling, M.D., Department of Surgery, University of Wisconsin, Hospital and Clinics–G5/348, 600 Highland Ave., Madison, WI 53792.

"Termination of intractable hiccups with digital rectal massage," by M. Odeh, H. Bassan, and H. Oliven, *J Int Med* 1990; 227:143–146. (*JIR* thanks Robert S. Hoffman for bringing this to our attention.)

abdomen when it is required on an almost daily basis." He adds that he uses Talon zippers which are "less likely to disengage and lead to evisceration, in comparison to YKK zippers" (see above).

As the French say: Reality depasses la fiction.

1. Transactions of the American Society for Artificial Internal Organs, Los Angeles, 29, 1956.
2. Freeman L. *Fight against Fears.* New York: Crown 1951.

Styles, trends, and tidbits culled from leading research journals
by Alice Shirell Kaswell

When you mix Erno Lazlo with water, there is nothing like it, according to a report on page 55 of the February 1991 issue of the research journal *Mirabella*. The report finds that Erno Laszlo Heavy Normalizer Shake-It is good science and good sense.

Naughty Nakeds

A clear sequin overlay can be used to construct a tank dress, according to a military science report on page 4 of the same journal. A related report (pp. 80–86) by investigator Nancy Collins tells how men in Hollywood assembled a loose cannon by using Julia Phillips as construction material. Calvin Klein has apparently extended the theoretical work begun by physicist Stephen Hawking. Klein's investigations into time—and Eternity in particular—are described on pp. 10–11. A report on pp. 16–22 tells about Ultima II's recent work on the Nakeds, skin-toned tones, bare-to-brown lipchromes, the Naughty Nakeds, and the relationship between artificial colors and artificial-looking colors.

Return of the Beautiful Mess

Estée Lauder continues to produce astonishing results in the fields of applied mathematics and computer science. *Mirabella*'s report (pp. 18–19) reveals how Lauder's Time Zone Moisture Recharging Complex can reprogram skin. The report also reiterates how Time Zone Eyes is a totally unique formulation called a liquicreme and emphasizes that it is an ultra-hydrating complex. Anthropologist Julie Baumgold finds, in first-person accounts on p. 20 and pp. 28–29, that 14% of her life is glamorous. Baumgold also concludes that Western civilization will return, again and again and again, to people who are beautiful, make messes, and lose things.

A Vibrating Man

Mirabella also presents several important psychology reports. Investigator Abigail McGanney examines (p. 40) the case of identical twins Jackie and Pauline, who both trained as psychiatric nurses before impulsively joining a London dance band. Investigator Donald Newlove describes (pp. 42–43) the case of a man who could pour seltzer on his head and smash eggs on himself. Investigator Ross Wetzsteon presents (p. 43) the case of a young man who focuses energy like a spotlight, becoming a vibrating presence. On page 47 there is a technical report about a moving image of a trapeze dress.

Femininity, a Vat, Teeth

Oscar de la Renta experiences the power of femininity, according to *Mirabella*'s puzzling yet fascinating report on page 79. Investigator Gene Stone reports (pp. 117–119) on the American Aroma Therapy Association (AATA) Convention and Trade Show. Aroma therapy, Stone reminds us, was created in 1937 when a French research chemist named Rene-Maurice Gattefosse placed his burned hand in a vat of lavender oil. At the AATA convention, Stone learned that unless he eats less tuna fish, rubs his ankles, and gets married, in 12 more years he will have trouble in his you-know. According to a report on pp. 122–123, with EpiSmile's unique Measuring Device one can actually see one's teeth becoming whiter.

Happy Hour Corpus

A strange medical/horological report on pp. 158–161 indicates that Happy Hour has bled into dinner time. The great thing about Italy, *Mirabella*'s investigator Frederick Eberstadt concludes on page 178, is that you can feed the corpus as well as the animus.

Tide and a Sunset

An intelligent laundry detergent called "Tide" is described on page 7 of the Spring 1991 issue of the research journal *Connections*. The report explores what happens if the detergent itself is required to be clean. A treatise by economist Mary Rowland (p. 12) discusses a formula that financial planners call the "Rule of 72." Rowland concludes that the formula is called the "Rule of 72" because it requires dividing the number 72 by another number. A meteorological report on page 22 finds that Brad Marx of the University of California, Santa Barbara, has preserved a lingering sunset.

Atomic Balls

Atomic golf balls are the subject of a report on page 4 of the vol. 22, no. 4 (April 1991) issue of the research journal *Executive Edge.* Low levels of radiation are used to rearrange molecules within the golf balls to make them bounce better and travel farther. *Executive Edge* says that by calling (204) 753-2311 you can arrange to have this done to your balls, free of charge.

Durable Minceur

Cellulite relief is examined on pp. 18–19 of section 6 of the May 12, 1991, issue of the research journal the *New York Times*. The *Times* discusses the probability that cellulite is a woman's birthright. The Skin-Care Laboratories of Lancôme (Paris) have conceived another breakthrough, this time involving a battery of anti-cellulite specifics, a targeted delivery system, micro-transport time-released action, and a modern mix of botanicals. The breakthrough, known as Durable Minceur Cellulite Relief Gel, is from France. The *Times* says that cellulite care is a sophisticated tradition in France. A report on page 5 describes a mascara formula, developed by Estée Lauder, that is more than lengthening, more than thickening, more than separating. Clinique Laboratories' latest astounding discovery, M Lotion, is the subject of a terse report on page 13. M Lotion is unscented, non-greasy. M Lotion makes no fuss.

Funny Money

Investigator G.M. Turk presents a salient first-hand analysis on page 19 of the February 1991 issue of the economics research journal *Money Making Opportunities*. Turk reports that they all laughed when he said he was going to start his own business. He presents proof of this.

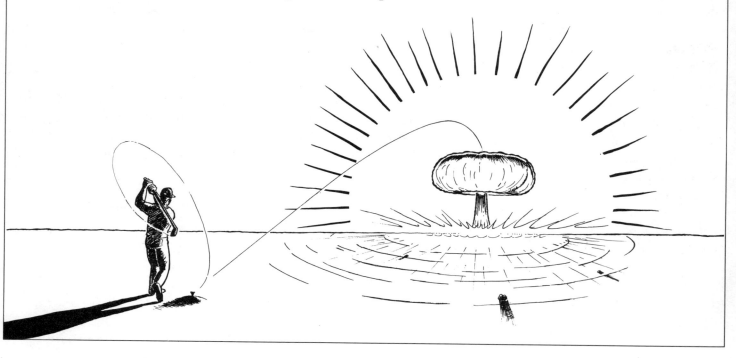

Maurice Kessler

CHAPTER 15

OVERLY STIMULATING RESEARCH

THE GLASS AND SPLEEN EXPLOSIONS

Sidney L. Saltzstein, M.D.

San Diego County General Hospital, San Diego, California

"Explosions" and their dire consequences occupy much of the scientific and popular literature. The world's population is growing at such a rate that if it were maintained until the year 6000, a solid mass of humanity would be expanding outward from the earth at the speed of light.[1] The number of authors per scientific paper is increasing so rapidly that within 15 years it will reach infinity.[2] Another pair of phenomena, possibly more to be feared, is the glass and spleen explosions.

Materials and Methods

When the surgical pathology laboratory at Barnes Hospital in St. Louis, Missouri, was remodeled in 1959, the number of specimens received annually in the laboratory was tabulated so that storage space for past and future slides could be planned. At the same time it was necessary to count and review the slides of all spleens either biopsied or removed and received in the same laboratory since the hospital was opened.[3]

Results

The annual number of surgical specimens received is plotted in Figure 1 on semilogarithmic coordinates. It is apparent that the number of specimens is increasing logarithmically (exponentially). Close examination of the graph also shows the effect of the Depression of the late 1920s and early 1930s, and of World War II. Determining the line of best fit by the least squares method, the following linear regression equation is obtained.[4]

$$\text{Log } 10N_1 = 2.175 + 0.0317 (T - 1900)$$

where N_1 is the number of specimens received in the year T. The coefficient of determination, correcting for degrees of freedom lost, is 0.916, indicating that 91.6% of the variation in the number of specimens received is explained by time alone. A "doubling time" of 9.49 years can be obtained from this equation. Extrapolating backwards, the year when the first specimen should have been received (the year when $N_1 = 1$) is 1832. While this may be absurd (Barnes Hospital did not open until 1912), Dr. William Beaumont, the first modern surgeon in St. Louis, arrived in the area in 1834 (an error of only 0.1%).[5] Perhaps the onset of this growth of surgical specimens should be correlated with his arrival.

Figure 1

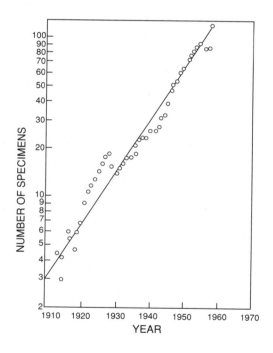

Extrapolating forward, in the year 2224, 2.854×10^{12} specimens will be received, making a total of 5.7×10^{12} specimens to be stored. The slides from approximately 1,119 specimens can be stored per cubic foot,[6] so 5.1×10^9 cubic feet of storage space will be needed. As the area of the city is 61 square miles,[7] this will be enough slides to bury the entire city of St. Louis under 3 feet of glass!

The annual number of spleens received is plotted in Figure 2, again on semilogarithmic coordinates.[8]

Again, it is apparent that the number of spleens is increasing logarithmically (exponentially), and the linear regression equation turns out to be:

$$\text{Log } 10N_2 = 0.917 + 0.0457 \ (T - 1900)$$

where N_2 is the number of spleens received in the year T. The coefficient of determination, correcting for degrees of freedom lost, is 0.773, indicating that 77.3% of the variation in the number of spleens received is explained by time alone. A "doubling time" of 6.59 years can be obtained showing that the number of spleens received is increasing considerably more rapidly than the total number of specimens received.

Figure 2

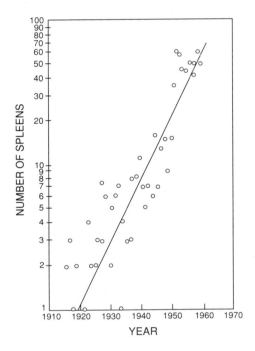

If one then sets $N_1 = N_2$, and solves the equation for both N and T, one will find that in the year 2121, 8,271,800,000 specimens will be received by the surgical pathology laboratory at Barnes Hospital, and all of them will be spleens!

Conclusion

Be careful when extrapolating biologic data.

1. Editorial. The weight of humanity. *St. Louis Post-Dispatch*, 1962; Sunday, July 22.
2. Price DJ de S. *Little Science, Big Science*. Cited in Review. *JIR* 1964; 13:22.
3. Saltzstein SL. Phospholipid accumulation in histiocytes of splenic pulp associated with thrombocytopenic purpura. *Blood* 1961; 18:73–88.
4. Hirsch WZ *Introduction to Modern Statistics*. New York: Macmillan, 1957.
5. Pitcock C DeH. The involvement of William Beaumont, M.D. in a medical-

50 AND 100 YEARS AGO IN *JIR*
50 Years Ago

August 1941: "Eyes of the scientific world widened this week at the startling news that work of a 27-year-old Columbia University graduate student has rendered null the notion that all matter is composed of atoms. The physicist is Dr. Elrod Hibbird. Hibbird's research indicates that everything men associate with the physical world, including atoms, molecules, our bodies and our mind patterns, is also illusory. Dr. Hibbird is currently engaged in preparing a book on the subject."

August 1941: "German soldiers are carrying small crystal pyramids in their pockets as they march into battle. Herr General Schwartzkopf has done research to demonstrate that a pyramidally shaped object is an inherently potent energy source."

100 Years Ago

August 1891: "Oat bran, fed to horses, leads to better wind."

August 1891: "Magic lantern video-graphic games are a pernicious influence on the young. They must be stopped."

August 1891: "A gentleman in New York City, an acquaintance of the mayor, is starting a hair club, for men who have lost their natural allotment."

August 1891: "They call it the economic theorem deriving from the supplier side, and it was proposed by Professor Gelder of the Commercial Correspondence Institution of Topeka. A reactionary idea of this sort could have originated only in the farther mentalist reaches of the West."

August 1891: "A photographic machine for copying documents in astonishing detail was demonstrated at the Jubilee Hall. Only one minor technical problem remains to be solved before the machine can be useful in the nation's commerce; the mechanical apparatus that manipulates the sheets of paper currently is prone to breaking down with great frequency."

legal controversy: the Darnes-Davis case, 1840. *Miss Histor* Rev 1964; 31–45.
6. Saltzstein SL. Unpublished data.
7. *World Almanac*. 1960; 294.
8. In accordance with Furst's first modification of the scientific method, the years when no spleens were received ($\text{Log}_{10}0$ indeterminant) are ignored.[9]
9. Furst A. On the treatment of annoying but incontrovertible and inexplicable facts. *JIR* 1964; 13:10.

A FIELD EXPEDIENT FOR MEASURING AMBIENT TEMPERATURE

Douglas U. Doubledome, M.D.

Director, Field Investigations,
Practicum Medical Center, Costa Mesa, California

In field work, traversing rough terrain and subject to environmental hazards, thermometers are constantly subjected to a high degree of attrition through loss and breakage. Since ambient air temperature is of importance to many disciplines and in management of camp environments, this attrition causes loss of essential data, inefficiency, and unnecessary discomfort.

After many years of suffering the results of these losses, our laboratory has devised an expedient that cannot be lost or broken. The method that we have devised requires calibration only once every 5 years (in most cases longer) or in case of severe damage.

Methods

In a multidisciplined approach, our engineering, physiology, and logistic experts examined the resources that were always available in the field. Not surprising, but seldom considered, were the human investigators themselves. We looked at all aspects of human capability to determine how these might relate to temperature. In this case it is unfortunate that humans are warm-blooded; insects and reptiles vary their rate of metabolic activity according to the ambient temperature. We did discover one aspect of the humans that responded dependably to the environment.

The male human testes require a temperature lower than the human body constant for production of quality sperm. To effect this, the scrotum sags down at higher temperatures, to expose more surface area, and pulls up close to the body at cooler temperatures. We hoped that this might be a way to objectively measure tem-

Figure 1. Distribution of *K* factor.

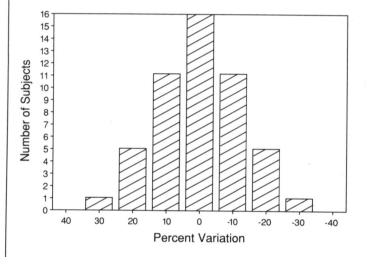

perature. It was realized at the outset that each male would most likely have to be individually calibrated, and we therefore obtained 50 test subjects from our student body.

Small thermocouple beads on extension rods were attached to the bodies of the test subjects on all available sides of the scrotum to ensure accurate measurement of the ambient temperature. Additional thermocouples were mounted on stands in the vicinity of the test subjects to ensure that heat radiated from the test subject's body did not skew the data. Although early in the experiment it was found that these were unnecessary, their use was continued to ensure that all data were presented in the same manner.

Data from each test subject was calibrated 10 times at different times of day and on different days of the week, over a period of 2 months. It was found that the average subject's scrotum extended 1 inch from 32° F to 72° F. This was, of course, not linear because the function of the extension was to expose more surface area. The function closely follows the square law, modified only by the convolutions of the individual scrotal surface. As a first approximation, the following relationship applies:

$$\Delta I = \sqrt{\Delta T}$$

where ΔI is the change in scrotal length and ΔT is the temperature in degrees Celsius. The K factor is an individual correction factor for differences in individual test subjects. The K factor varied in the experimental population according to the classical Gaussian (bell-shaped) curve. This was somewhat surprising in a population as small as 50 subjects, but gave us renewed confidence in the validity of the methods used. Figure 1 shows this distribution.

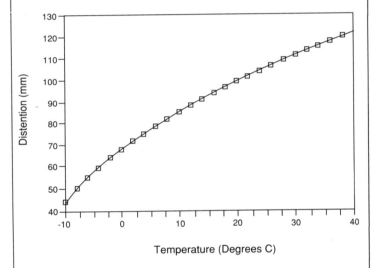

Figure 2. Distention vs. temperature.

Figure 2 shows the relationship of scrotal length to temperature for test subjects in the median range ($K = 1$). We were surprised at how rapid the reaction was, taking less than 60 seconds to stabilize a new temperature.

In our first model, we were influenced by the beam galvanometer (shown in Figure 3). While this device yielded practical results, it was later simplified and improved.

The shadow of the left portion of the scrotum is cast on an individual calibration strip on the left leg. With the light source in the optimum position, the advantage of a source-to-scrotum distance of about 3 inches and a source-to-calibration distance of 9 inches gives

a multiplier of 3 to the scrotal deviation.

This method worked well in the laboratory, but it was soon realized that it required test subjects to be in subdued light and required a light source that could be as easily lost or damaged as a thermometer. Attempts to use alternative light sources such as candles or torches were flawed by the radiated heat from the alternate source warming the scrotum and skewing the results and by some unfortunate accidents which reduced our pool of test subjects.

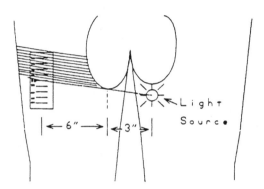

Figure 3. First test device.

Once the principle was established and proved workable, we improved the device to eliminate the light source and to eliminate any equipment but the calibrated human body. The second device is shown in Figure 4.

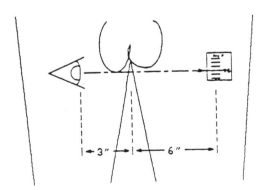

Figure 4. Improved model.

In this version we have shifted the fulcrum point to the left side, since the part of the scrotum containing the left testicle hangs lower, and this improves the sen-

THE ERADICTION OF POVERTY IN CONTEMPORARY AMERICA: GENESIS, TREATMENT AND EVALUATION[1]

Alan Frankel[2]
University of Portland, Portland, Oregon

Abstract

Without explicating specific psychologic, social, political, and economic theories, and utilizing a hypothetico-deductive epistemology, a testable theorem is derived concerning the ontogenesis of poverty in contemporary America. A treatment strategy is suggested as well as a theory-related method of evaluation.

Poor people do not have enough money. Give them enough money. They will no longer be poor.

Figure 1. Mathematic representation of the Poverty-Money relationship ($r = 1.00$, $p < .0000001$).

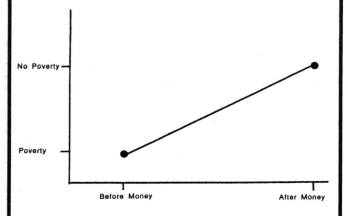

---

[1] This study was financed in part by Grant No. RD 14-1063-7093/RPM-1049/CIA/OSS/POOF from the Society for the Preservation of Social Sciences Interested in Nonparametric Hyper-multivariate Research of Trivial Social Problems.

[2] Reprints are available from the author at the Department of Psychology, University of Portland, 5000 N. Williamette Boulevard, Portland, Oregon 97203, if you are absurd enough to request one.

sitivity of the reading. The assistant's eye is positioned adjacent to the marked point on the left thigh, and by sighting across the scrotum to the calibrated scale on the right leg, a small object (a pencil point, a ball-point pen point, needle or pin, etc.) is placed on the calibration scale in line with the lowest portion of the scrotum. While this object is held in place, the assistant can move around to read the scale.

This method has the advantage of requiring only the calibrated human body and some writing instrument, essential to field investigations. The marked point on the left thigh and the calibrated scale on the right thigh were tattooed on the test subjects.

REFERENCES

Hancar F. Zum Problem der Venusstatuettenim Eurasiatischen Jungpalaolithikum. *Praehistorische Zeitschrift* 1978; 45:3002.

Gray H. *Anatomy of the Human Body* 1973; 1923.

Loeb F. Harnrohrencapacitat und Tripperspritzen. *Munchen Med Wschr* 1956; 20:312.

Scott TP. Male Anatomy. *Chippingdale Annual* 1984; 10–199.

CHAPTER 16

THE OTHER NOBEL

NO DEATHS AT FIRST IG NOBEL CEREMONY

Overflow Crowd Witnesses Shocking Incidents

Althea P. Plovkcic
Society for Basic Irreproducible Research

Photographs by Roland Sharrillo

Circumstance, pomp, delight; dismay, shocks, and disruptions, music, mayhem, and a whiff of scandal were the order of the evening at the First Annual Ig Nobel Prize Ceremony, held at the MIT Museum in Cambridge, Massachusetts, on the evening of October 3, 1991.

An overflow crowd of 350 dignitaries and well-wishers (and one angry woman) watched as 10 worthy individuals stepped forth to receive Ig Nobel Prizes. The Prizes were presented by (non-Ig) Nobel Laureates Eric Chivian (Peace, 1985) of Harvard University and MIT, Henry Kendall (Physics, 1990) of MIT, Sheldon Glashow (Physics, 1979) and Dudley Herschbach (Chemistry, 1986), both of Harvard.

A fifth Laureate, Jerome Friedman (Physics, 1990) of MIT, appeared in the form of a slide projection and several tape-recorded speeches. Friedman, who was in Japan at the time of the ceremony, expressed his hope that the audience was "enjoying this as much as I am." Later in the evening, a tape recording delivered his personal message to each of the new Ig Nobel Laureates: "Congratulations. Your work is an inspiration to all of us."

Renegade Laureates

The (non-Ig) Nobel Laureates continually attempted to subvert the dignity of the proceedings. As the ceremony was beginning, they all put on Groucho glasses (plastic glasses with attached false probosci and mustaches) and then, in concert with a ceremonial flourish from the brass quintet, simultaneously crossed their legs. All four men, especially Professors Chivian and Kendall, shamelessly winked, gestured, and smiled at their many admirers in the audience.

The Swedish Meatball King was sufficiently upset by their behavior that, in the midst of the ceremony, he commanded the Master of Ceremonies to request that the Laureates refrain from making eyes at the Swedish Meatball Queen. With the exception of Professor Glashow, they did.

Glashow was a one-man wrecking crew, repeatedly disrupting the ceremony. Seating himself next to the speaker's podium and wielding a black cane, he poked speakers in the back, pushed the Master of Ceremonies' hat off from the rear, and on one occasion went so far as to hook a speaker (T.F. Bakker) around the neck and bodily remove her from the stage. Throughout, he could not be prevented from heckling any dignitaries and other audience members who were unfortunate enough to attract his attention.

Some measure of discipline was maintained by the Umpire, John Barrett, a man of stentorian voice and authoritative manner, who was equipped with a face mask and a chest protector.

Disruptive Visitors and Voices

Barrett could exercise only limited control, however, over the many unexpected visitors who ran, rolled, cartwheeled, bashed, bicycled, and wandered their way into the hall after the ceremony had begun.

Three times, an angry young woman in a red dress strode into the hall. Dragging a policeman by the arm,

Over 100 important personages, all wearing numbered badges ("DIGNITARY # 43"), entered the hall to cheers and huzzahs. Clad in gowns, lab coats, naval uniforms, bathrobes, wedding gowns, feathers, sashes, cummerbunds, fright wigs, leather jackets, lederhosen, and other traditional raiments of academia, they marched through the aisles and to their seats.

The audience members, all wearing numbered badges ("AUDIENCE # 237"), applauded enthusiastically. Many were attired in lab coats, which took on many hues as confetti and excelsior wafted through the air and settled on them.

A torchbearer then entered the hall, cried "Excelsior," ran onto the stage, bowed to the Swedish Meatball King and Queen, bowed to the Laureates, and vanished.

she defiantly pointed to the Laureates seated onstage, shouting, "That's the man, officer! That's the father of my child!" (Professor Glashow, upon whose shirt was emblazoned the slogan "Nobel Stud," was uncharacteristically silent during these outbursts.)

Choral Director Stephen Baum, too, apparently felt impelled to inject himself into the proceedings. Whenever any of the speakers happened to utter the phrase "let us hail the unsung genius of . . . ," Baum, a baritone, rose from his seat and began singing the chorus from Walton's "Belshazzar's Feast."

"Welcome, Welcome"

The evening began with the ceremonial Parade of the Dignitaries, led by Prima Ballerina Sally Beddow (curator of the MIT Museum's photo archives), Master of Ceremonies Marc Abrahams (editor of **JIR**), and the Nobel Laureates.

He was to reappear sporadically throughout the evening, carrying lamps, candelabra, and other implements of illumination.

The traditional Ig Nobel Welcome, Welcome speech (ie, "Welcome, welcome") was followed by numerous lengthier orations and perorations, rites, and incantations. Kathleen Thurston, a descendant of Ignatius No-

*Left to right: **Nobel Laureates Dudley Herschbach (seated) and Henry Kendall; the Torchbearer; the Swedish Meatball Queen and King with Chief of Protocol Michèle Meagher; T.F. Bakker delivering the acceptance speech on behalf of Michael Milken as Sheldon Glashow applies his cane to her neck; angry woman leads a policeman toward the stage.***

HISTORY OF THE IG NOBEL PRIZE

The Ig Nobel Prizes are a legacy from the estate of the late Ignatius ("Ig") Nobel, believed to be the inventor of soda pop and the co-inventor of excelsior (packing material made from wood chips). Ig Nobel claimed to be a relative of Alfred Nobel, the man who invented dynamite and endowed the other Nobel Prizes. Ig Nobel shared the family fascination with things that go "pop." His careful observations of soda pop bubbles, conducted over a period of 17 years, led him to conclude that no two soda pop bubbles take the same path. This was Ig Nobel's greatest achievement; it has never been reproduced.

Ig Nobel once said, "An irreproducible achievement is one that either cannot be reproduced or shouldn't be." He specified, via his will, that his accumulated wealth be used to honor individuals for notably irreproducible achievements in science and other areas of human endeavor.

In 1991, after some decades of administrative delay, the first annual public ceremony was mounted. Special funding was provided by the Peter deFlorez '38 fund for humor at MIT.

bel's podiatrist, made her obligatory appearance. In keeping with the family tradition, she declined to speak.

Handing It to Them

Each new Ig Nobel Laureate delivered a brief speech and was then presented with a parchment and an Ig Nobel Prize. The Ig Nobel Prize is a pancake-shaped device that screams when shaken. It is surmounted by the Ig Nobel logo and suspended from a golden rope. The parchment itself is blank.

Each of these valued objects was passed directly from the Swedish Meatball Queen to the Swedish Meatball King to Chief of Protocol Michèle M. Meagher to Master of Ceremonies Abrahams, and then to Professor Glashow, to Professor Chivian, to Professor Herschbach, and to Professor Kendall. Professor Kendall, at his discretion, then either presented it directly to the new Ig Nobel Laureate or handed it back to one of his Nobel colleagues, who then made the presentation. Sev-

eral of the Ig Nobel Laureates became disoriented by the process.

Each of the new Ig Nobel Laureates was also presented with a parking pass, valid between 3 A.M. and 4 A.M. on Christmas Day, from the Cambridge Chamber of Commerce. They were also offered free career counseling from Steybel, Peabody, Lincolnshire, the fourth largest outplacement firm in the United States.

Accepting the Consequences

The acceptance speeches sparkled with surprises and circumlocution. Thomas Kyle, the 1991 Ig Nobel Physics Laureate, distinguished himself by becoming the first person to refuse an Ig Nobel Prize. Kyle explained that he did not make the discovery or do the research that earned him the prize. "I just mailed the paper in," he said. (Kyle's article about the discovery of administratium, the heaviest known element—or to be more accurate, an article with Kyle's name attached to it—appeared in *JIR* 35:1.)

Paul Defanti, inventor of the Buckybonnet and winner of the Prize for Pedestrian Technology, took the opportunity to implore the crowd to purchase his device, and was physically removed from the platform.

The Education Prizewinner, J. Danforth ("Dan") Quayle, was honored "for demonstrating, better than anyone else, the need for science education." Looking much younger and shorter than he usually appears on television, Quayle delivered a stirring speech, which is here reproduced in full:

"Thank you. Thank you for this great honor. As you know, education is very important to me. I came to value it rather late in life. When I was in college, I wasted my time drinking beer, playing golf, and chasing babes. Now that I am Chairman of the National Space Council and Vice President of the United States, I sorely regret my lack of knowledge about science and technology. Tomorrow, I am going to return to Washington and urge the President to join me in taking up the study of mathematics. I look forward with great joy to learning the calculus. Soon I will be able to take derivatives of simple continuous functions. Then I will learn to do integrals. I will understand the relationship between displacement, velocity, and acceleration. I believe every public officeholder should have these skills. Thank you."

The Chemistry prizewinner, Jacques Benveniste, whose iconoclastic paper about water, memory, and homeopathy in the research journal *Nature* won him the prize, did not attend the ceremony. His acceptance speech was delivered by Kurt Hasselbalch, the gruff-spoken Curator of the Hart Nautical Collection at the MIT Museum and Admiral of the Great Navy of the State of Nebraska. Hasselbalch, who drank several

Robert Klark Graham, the entrepreneur whose "Nobel sperm bank" earned him the Biology Prize, was

HOW THE WINNERS WERE CHOSEN

The Ig Nobel Prize Nominating Committee received nominations from scientists and distinguished citizens of 37 nations. The final decision was made by a collaborative stochastic process involving the editors of *JIR*, a select group of MIT and Harvard faculty members, the (non-Ig) Nobel Laureates, the Ig Nobel Posthumous Board of Governors, and four or five people who were stopped on the street.

also not present. Accepting on his behalf was Dr. Stanley Strangelove, a well-known early advocate of sperm-banking. Dr. Strangelove decried the fact that none of the Nobel Laureates seated on the dais had yet contributed to Graham's enterprise.

The Literature Prizewinner, Erich Von Daniken, was not in the room at the moment his prize was presented. Robert Richard Smith, the manager of a local softball team and an admirer of Von Daniken's methodology, graciously stepped to the podium and delivered a rambling acceptance speech on Von Daniken's behalf.

glasses of water prior to the ceremony, forgot most of what he had planned to say.

On the day of the ceremony, Josiah Carberry, Brown University's legendary professor of psychoceramics (the study of cracked pots), was on a research expedition in search of the Amazon River. A delegation from Brown, led by University Librarian Merrilee Taylor, accepted the Interdisciplinary Research Prize on Carberry's behalf.

Left to right: *The projected photo of Nobel Laureate Jerome Friedman (photo: Robert Rose); Master of Ceremonies Marc Abrahams with T.F. Bakker; Nobel laureates (from left) Sheldon Glashow, Eric Chivian, Dudley Herschbach, and Henry Kendall. Vice President Dan Quayle delivers his acceptance speech (photo: Michelle Green, The Tech), Nobel Laureate Sheldon Glashow and Ig Nobel Laureate Alan Kligerman.*

Edward Teller, the Ig Nobel Peace Prize winner, was not present. The Ig Nobel Committee was unable to find anyone to accept on his behalf.

Economics Laureate Michael Milken, the inventor of the junk bond, was unable to attend the ceremony. Unfortunately, Mr. Milken had a previous 15- to 20-year engagement.

A speech on his behalf was delivered forcefully, tearfully, and at great length by T.F. Bakker, who explained that she had come to know Mr. Milken recently when he became a close colleague of her husband Jim, who also has a 15- to 20-year engagement. Ms. Bakker's reminiscences were abruptly terminated by Professor Glashow and his cane.

The final award of the evening was presented to Alan Kligerman. Kligerman won the Ig Nobel Medicine Prize for his invention of Beano, a substance that prevents the gas associated with beans, cabbage, broccoli, and many other hard-to-digest foods. Kligerman's buoyant

enthusiasm raised clouds of enthusiasm throughout the hall. The high point of the ceremony came when he presented Beano samples to the four Nobel Laureates and instructed them in its use.

Speculation among the public and the press had handicapped the cold fusion researchers as odds-on favorites to cop an Ig Nobel Prize. Master of Ceremonies Abrahams explained that the Ig Nobel Committee had chosen not to award an Ig Nobel Prize this year to the discoverers of cold fusion. "The committee felt that it is simply too soon to tell whether cold fusion is irreproducible," Abrahams announced. "That is something that will become apparent in the fullness of time."

"Goodbye, Goodbye"

The evening drew to a close with the traditional Ig Nobel Goodbye, Goodbye speech ("Goodbye, goodbye").

The torchbearer vanished from the hall, and the 1991 Ig Nobel Laureates strode boldly off history's stage.

THE 1991 IG NOBEL PRIZEWINNERS

The 1991 Ig Nobel Physics Prize. Thomas Kyle, detector of atoms and original man of knowledge, for his discovery of the heaviest element in the universe, administratium.

The 1991 Ig Nobel Pedestrian Technology Prize. Paul Defanti, wizard of structures and crusader for public safety, for his invention of the Buckybonnet, a geodesic fashion structure that pedestrians wear to protect their heads and preserve their composure.

The 1991 Ig Nobel Education Prize. J. Danforth Quayle, consumer of time and occupier of space, for demonstrating, better than anyone else, the need for science education. (The acceptance speech was delivered by someone who claimed to be Vice President Quayle but who appeared to be an 8-year-old girl.)

The 1991 Ig Nobel Chemistry Prize. Jacques Benveniste, prolific proselytizer and dedicated correspondent of **Nature,** for his persistent discovery that water, H_2O, is an intelligent liquid and for demonstrating to his satisfaction that water is able to remember events long after all trace of those events has vanished.

The 1991 Ig Nobel Interdisciplinary Research Prize. Josiah Carberry, bold explorer and eclectic seeker of knowledge, for his pioneering work in the field of psychoceramics, the study of cracked pots.

The 1991 Ig Nobel Biology Prize. Robert Klark Graham, selector of seeds and prophet of propagation, for his pioneering development of the Repository for Germinal Choice, a sperm bank that accepts donations from Nobellians and Olympians.

The 1991 Ig Nobel Literature Prize. Erich Von Daniken, visionary raconteur and author of Chariots of the Gods, for explaining how human civilization was influenced by ancient astronauts from outer space.

The 1991 Ig Nobel Peace Prize. Edward Teller, father of the hydrogen bomb and first champion of the Star Wars weapons system, for his lifelong efforts to change the meaning of peace as we know it.

The 1991 Ig Nobel Economics Prize. Michael Milken, titan of Wall Street and father of the junk bond, to whom the world is indebted.

The 1991 Ig Nobel Medicine Prize. Alan Kligerman, deviser of digestive deliverance, vanquisher of vapor, and inventor of Beano, for his pioneering work with antigas liquids that prevent bloat, gassiness, discomfort, and embarrassment.

Newly elected member of the Posthumous Board of Governors. Rube Goldberg.

Newly Elected Honorary Member of the Posthumous Board of Governors. Marilyn vos Savant.

The Ig Nobel logo was designed by Lois Malone/Rich & Famous Graphics. It reflects traumatic themes from the artist's childhood, and incorporates images from the life of Ignatius Nobel.

Clockwise from top left: Prima Ballerina Sally Beddow (left) uses a glass of water to explain the Benveniste chemistry experiment to one of the dignitaries; Warren Seamans, director of the MIT Museum (rear); a face in the crowd; Torchbearer at rest; group photo (front) Barbara Linden, the Ig Nobel Authority Figure, Kurt Hasselbalch, Dr. Stanley Strangelove, Paul Defanti in his Buckybonnet, (rear) Nobel Laureates Glashow, Herschbach and Kendall, and JIR's Alice Shirell Kaswell.

TEN HONORED AT SECOND FIRST ANNUAL IG NOBEL PRIZE CEREMONY

Stephen Drew and Francesca Thurston

This year's winners of the Ig Nobel Prize were honored on October 1 at a ceremony held at MIT's Kresge Auditorium and sponsored jointly by the *Journal of Irreproducible Results*, The MIT Museum and Kelvin,™ the official fragrance of the 1992 Ig Nobel Ceremony. The prizes honor individuals whose achievements cannot or should not be reproduced. Approximately 800 people, many of them wearing lab coats, antique military uniforms, bathrobes or—in at least two cases—gorilla suits, viewed the proceedings.

This year's ten Ig Nobel laureates come from five different countries: Japan, France, England, Russia, and the United States. The winners were nominated by an international committee of distinguished scientists and pedestrians—but in truth, the winners selected themselves. They were presented with the Ig Nobel Prize, a cheap painted hand mirror that screams when shaken. Each winner also received a gift from the Cambridge Chamber of Commerce: a map that shows the secret locations of the three public restrooms in Cambridge.

Amidst pomp and circumstance, and with continual interruptions from hecklers, singers, protesters, weather balloons, a bicycle messenger, a dog, an indoor blimp and an Olympic torchbearer, the ceremony's participants included genuine Nobel laureates Sheldon Glashow of Harvard (Physics, 1979), Jerome Friedman of MIT (Physics, 1990), and Mel Schwartz of Brookhaven National Laboratory (Physics, 1985). Glashow was biologically present. Friedman and Schwartz appeared with the aid of electronic and photographic technology. The blimp and weather balloons were provided by Kelvin™, the official fragrance of the 1992 Ig Nobel Prize ceremony.

THE TRADITIONAL WELCOME, WELCOME SPEECH

The traditional Ig Nobel Welcome, Welcome speech was delivered by Barbara Linden of the MIT Museum. Below is a complete transcript of the speech:

Welcome, welcome.

Photo courtesy of Roland Sharrillo

The Swedish Meatball King presided with quiet grace and dignity, seemingly unperturbed by the catcalls of hecklers who pointed out that the woman at his side was "not the same Queen he had last year." Spokesmen for his Highness declined to comment on the scandal which has been unfolding noisily in recent months in the pages of the European tabloid newspapers. Seat cushions for the King and Queen were sponsored by Kelvin™, the official fragrance of the 1992 Ig Nobel Prize ceremony.

searchers wearing silver flame-retardant suits worked feverishly at the back of the stage, assembling a melange of laboratory apparatus. The trio, Astrid Hiemer, Joe Davis, and Binya Kessaly, were amassing data for a paper they submitted last May to the research journal **Cell**. *JIR*'s Elegant Results columnist, Alice Shirell Kaswell, who attempted to monitor the researchers, later said that aside from campaigning to win their own Ig Nobel Prize, Hiemer, Davis, and Kessaly seemed to have in mind three entirely unrelated experiments. The laboratory equipment was donated by Kelvin,™ the official fragrance of the 1992 Ig Nobel Prize ceremony.

Music was composed and performed by world-renowned jazz harpist and singer Deborah Henson-Conant. Henson-Conant, who is perhaps best known to *JIR* readers as the curator of the Museum of Burnt Food (see "A Decade of Burnt Food," page 22), also performed a rousing duet on cymbals with Sheldon Glashow. A second harpist, Lois Malone, spent the evening onstage reading a book and grooming her nails, hair and makeup. Ms. Malone's appearance was underwritten by Kelvin™, the official fragrance of the 1992 Ig Nobel Prize ceremony.

The Bolzmann Consort, a physics brass quintet based at MIT, lent tone, melody and tempo to the proceedings, as did the giant pipe organ played by Nathan Tater Abrahamson. As at last year's ceremony, itinerant opera singer Stephen Baum was in the audience and could not be restrained from breaking into song whenever he heard the phrase "unsung genius."

Throughout the evening, three untenured MIT re-

Left to right: Ivette Bassa, the inventor of bright blue Jell-O®, accepts the chemistry Prize. Paying homage are Nobel physics laureate Sheldon Glashow (wearing white headgear), the Swedish Meatball Queen and the Swedish Meatball King. Professor Glashow's headgear was provided by Kelvin™, the official fragrance of the 1992 Ig Nobel Prize ceremony (photo: Luke D'Ancona). Human spotlights illuminated the evening. The spotlights were sponsored by Kelvin™, the official fragrance of the 1992 Ig Nobel Prize ceremony (photo: Dr. Robert Richard Smith/Minimum Wage Art). The audience used large and small weather balloons to engage in a symbolic debate on the merits of "Big Science vs. Little Science." Little science (not shown) won the day (photo: Roland Sharrillo). The Swedish Meatball Queen (left) and the Swedish Meatball King (right) acknowledge the legacy of Ignatius Nobel (photo: Ben Wen). David Vlach (wearing the tie and business suit) of the Cambridge Chamber of Commerce offers gifts to the new Ig Nobel laureates. Vlach also emotionally paid tribute to the Kelvin Perfume Company and its role in revitalizing the local economy. At left is Dr. Jacques Boisse, a representative from the Kelvin Company (photo: Luke D'Ancona).

The ceremony was briefly supplanted by a spontaneous Buckybonnet fashion show organized by Paul DeFanti, the 1991 Ig Nobel Prizewinner for Pedestrian Technology, and was narrated by fashion celebrity Lisa Mullins. Ms. Mullins' gown was provided by Kelvin™, the official fragrance of the 1992 Ig Nobel Prize ceremony.

Just prior to the announcement of the final Prize, a bicycle messenger rode into the hall and took the stage to read a telegram from the International Olympic Committee decrying the "blatant commercialism of this year's Ig Nobel ceremony."

The evening was supported in part by a grant from the Peter C. DeFlorez '38 Fund for Humor at MIT. Generous support was also provided by Kelvin™, the official fragrance of the 1992 Ig Nobel Prize ceremony, by Rob Falk Productions, the official lighting design specialists of the 1992 Ig Nobel Prize ceremony, and by Stinsen Bicycle Messengers, the official delivery service of the 1992 Ig Nobel Prize ceremony.

The crowd was well behaved, with the exception of MIT materials science professor Robert Rose, who was expelled from the balcony for directing paper airplanes and spitballs at the royal couple.

The Ig Nobel Speeches Highlights

Biology: (Unfortunately Dr. Jacobson could not attend the ceremony. He had a previous 15- to 20-year engagement. His award was accepted by Bob Cecil Turner and Larry Cecil Wilson.)

Larry: We have never met that prolific man who we

HISTORY OF THE IG NOBEL PRIZES

Ig Nobel Prizes are awarded to individuals whose achievement cannot or should not be reproduced. The Prizes are a legacy from the estate of the legendary Ignatius ("Ig") Nobel, the inventor of excelsior (packing material) and co-inventor of soda pop. The first Ig Nobel Prizes were awarded by the ***Journal of Irreproducible Results*** in 1968 (for a detailed account see *JIR* 16:3, April 1968); the winners have been consigned to obscurity, where they will soon be joined by this year's laureates. The first public ceremony was held in 1991 (see page 168).

call . . . Dad. But there is a haunting familiarity about him and his work. You can see it throughout our neighborhood.

Bob: These days, folks say it's hard to understand each other because everyone comes from different backgrounds. But when you get down to it, we're all a lot more alike than we are different.

Chemistry: (The acceptance speech was given by Ig Nobel laureate Ivette Bassa.)

I feel humbled at being singled out for this honor. My achievement is simply the capstone of an immense body of scientific research performed over the past hundred years. My colleagues and I wish to direct your attention to three great chemists who laid the foundations for our work:

Experimenters Binya Kessaly (left), Astrid Hiemer and Joe Davis attempt to manufacture data (photo: Enzo Crivelli).

Emil Fischer, who won the Nobel Chemistry Prize in 1902 for synthesizing sugars and purine derivatives; Richard A. Zsigmondy who won the 1925 Nobel Prize for his pioneering work in colloidal chemistry; and Linus Pauling, who won the 1954 Nobel Prize for his discoveries about the nature of chemical bonds. Most of all, I would like to acknowledge our debt to the great 19th century scientist whose research opened the door to 20th century chemistry: Pearl B. Wait, a manufacturer of cough medicine in Leroy, New York, who in 1897 became the first person to synthesize Jell-O® gelatin dessert.

Physics: (The acceptance speech was given on behalf of David Chorley and Doug Bower by Frank Laughton of the Shave N' Spell Crop Circle Corporation.)

David Chorley and Doug Bower are heroes for creating all those crop circles in England. I've never met the men personally, but I say God bless 'em. I want to tell you about the crop circles in Brooklyn, but I don't have the time today, so here's my take on the international monetary crisis. As I see it, crop circles hold the key to the economic revitalization of Russia, the Baltic states and the other nations of Eastern Europe. Twenty years ago, Russia needed crops. Now they have plenty of crops, but no crop circles. That's where my company comes in. We plan to open three Shave N' Spell Crop Circle schools in Russia later this year.

We've got to retrain that work force. We've got to get the assembly lines back to full production. We've got to get back to the gold standard. And above all, we've got to prevent them from fluoridating the bottled water.

Peace: (The acceptance speech was given on behalf of Daryl Gates by Stan Goldberg, general manager of Crimson Tech Camera Store.)

As general manager of Crimson Tech Camera Store, I am pleased to accept this award on behalf of Daryl Gates. Daryl Gates has done more for the videocamera industry than any other individual. He has shown the world how a good quality videocamera can capture the memories of a generation. [Here, Mr. Goldberg held up a videocamera.] Take this baby, for instance. It's a Fuji with one-lux light sensitivity, an AF power zoom/macro lens, full-range autofocus, and automatic time-date stamping. We sell it for just $599.98, which includes a free bonus case. We'll beat any competitor's advertised price—[At this point several people rushed onstage, attacked Mr. Goldberg and removed him from the building. It is believed that a member of the audience

*Left to right: **Heisenberg harpist Lois Malone** (photo: Luke D'Ancona). **Buckybonnet fashions** were displayed at a spontaneous fashion show (photo: Roland Sharrillo). An audience member undergoes recalibration following the ceremony (photo: Enzo Crivelli). Half-siblings **Bob Cecil Turner** (left) and **Larry Cecil Wilson** (right) accept the Ig Nobel Medicine Prize on behalf of prolific sperm donor **Dr. Cecil Jacobson** (photo: Ben Wen).*

NEXT YEAR'S CEREMONY

Although the Prizes are given for irreproducible achievements, one aspect of the ceremony is fully reproducible: it is always held on the first Thursday evening in October.

If you have a delegation that wishes to attend, or if you desire further information, please contact the MIT Museum at (617) 253–4422.

---

videotaped the incident and is offering it for sale to the television networks.]

Nutrition: (The acceptance speech was given on behalf of the utilizers of Spam by war hero and lifelong Spam-eater Dr. Jack S. Meagher of Harvard University.)

Excuse me for eating with my mouth full. . . . I am proud, honored, and yes, hungry to accept this award on behalf of the thousands and thousands—nay billions and billions and billions and billions—of Spam eaters around the world, past, present and future. I have been eating this wonderful substance since 1942. I went off to the service and I was homesick. I was nothing until I found Spam in the Navy—and look at me now—ah, talk about family values and really, you're just talking about Spam.

Literature: (The acceptance speech was given on behalf of Yuri Struchkov by Yulia Govorushko, who delivered her talk in Russian. This is an approximate English translation.)

I don't know anything about publishing scientific papers, and I never heard of Yuri Struchkov. So instead, I will tell you how to pluck a chicken.

Take the feathers out. That's the important thing. You pull them out. Usually there are some left. You get them out by burning the chicken a little bit. You hold it above the flame for a while. The remaining feathers get loosened, and then you pull them off. That's it. You have plucked the chicken. Really, that's the whole story.

Anyone wishing more details, for God knows what reason, please see me later. I would prefer that you send me a letter, but I know better than to expect you would. No one writes letters anymore. Correspondence is a lost art. Plucking chickens is not the kind of thing that inspires people in the late 20th century to take up a pen and set thoughts to paper, at least in the West.

Where I grew up, the situation is different. People there care about chickens, because they are expensive. You people, to judge by your appearance do not invest your hopes and dreams in poultry. You believe you are free to choose your own destiny. Perhaps this is the case. I myself am unsure.

Left to right: Audience members used placards to communicate their views (photo: Enzo Crivelli). Stan Goldberg of Crimson Tech Camera Store accepts the Ig Noble Peace Prize on behalf of former Los Angeles Police Chief Daryl Gates. Moments after this photograph was taken, Mr. Goldberg was attacked and removed from the stage (photo: Ben Wen). Audience mem- bers show their appreciation of the new prizewinners' irreproducible achievements (photo: Luke D'Ancona). Jim Knowlton, co-winner of the Ig Nobel Prize for Art for his seminal work, "Penises of the Animal Kingdom" (photo: Luke D'Ancona.) Nobel Physics Laureate Sheldon Glashow (left) and acclaimed

THE 1992 IG NOBEL PRIZE WINNERS

Biology: Dr. Cecil Jacobson, relentlessly generous sperm donor, and prolific patriarch of sperm banking, for devising a simple, single-handed method of quality control.

Chemistry: Ivette Bassa, constructor of colorful colloids, for her role in the crowning achievement of twentieth century chemistry, the synthesis of bright blue Jell-O.®

Physics: David Chorley and Doug Bower, lions of low-energy physics, for their circular contributions to field theory based on the geometrical destruction of English crops, ie, crop circles.

Medicine: F. Kanda, E. Yagi, M. Fukuda, K. Nakajima, T. Ohta and O. Nakata of the Shisedo Research Center in Yokohama, for their pioneering research study "Elucidation of Chemical Compounds Responsible for Foot Malodour," especially for their conclusion that people who think they have foot odor do, and those who don't, don't. (Their paper was published in the British Journal of Dermatology, 122:771, 1990.)

Peace: Daryl Gates, former Police Chief of the City of Los Angeles, for his uniquely compelling methods of bringing people together.

Economics: The investors of Lloyds of London, heirs to 300 years of dull prudent management, for their bold attempt to insure disaster by refusing to pay for their company's losses.

Archaeology: Eclaireurs de France, the Protestant youth group whose name means "those who show the way," fresh-scrubbed removers of graffiti, for erasing the ancient paintings from the walls of the Meyrieres Cave near the French village of Brunquiel.

Nutrition: The utilizers of Spam, courageous consumers of canned comestibles, for 54 years of undiscriminating digestion.

Literature: Yuri Struchkov, unstoppable author from the Institute of Organoelemental Compounds in Moscow, for the 948 scientific papers he published between the years 1981 and 1990, averaging more than one every 3.9 days. (The statistics were compiled by David Pendlebury of the Institute for Scientific Information.)

Art: Presented jointly to Jim Knowlton, modern Renaissance man, for his classic anatomy poster "Penises of the Animal Kingdom," and to the U.S. National Endowment for the Arts for encouraging Mr. Knowlton to extend his work in the form of a pop-up book.

international jazz harpist Deborah Henson-Conant (right) premier their newly composed duet for cymbals. Professor Glashow's belt buckle was supplied by Kelvin™, the official fragrance of the 1992 Ig Nobel Prize ceremony (photo: Ben Wen).

Art: (The acceptance speech was given by Ig Nobel laureate Jim Knowlton.)

In this century, a rigid barrier has been erected between art and science. As both an artist and a scientist, I believe this to be a dangerous thing.

My seminal work is a comparative anatomy chart featuring the male copulatory organs of several animals from man to whale. I created the chart during my graduate studies at Columbia University. I have always been drawn to ambiguous and understated conceptual works. "Penises of the Animal Kingdom" exposes and exploits the tension between subject (the phallus) and dry context (the classical anatomy chart), stimulating artist and scientist alike to re-examine prevailing societal attitudes.

It is my hope and dream to develop a pop-up book version of "Penises of the Animal Kingdom." I have spoken at length with the National Endowment for the Arts. They have strongly encouraged me to apply to them for funding, and are advising me as to how they and I can best work together.

On behalf of art, and on behalf of science, and on behalf of the members of the animal kingdom, I thank you.

THE GOODBYE, GOODBYE SPEECH

The traditional Ig Nobel Goodbye, Goodbye speech was delivered by Barbara Linden of the MIT Museum. Below is a complete transcript of the speech: Goodbye, Goodbye.

Photo courtesy of Roland Sharrillo

RETURNING LAUREATE'S ADDRESS

This year's Returning Laureate Address was delivered by J. Danforth Quayle, Chairman of the National Space Council, who last year received the Ig Nobel Prize for Education. Mr. Quayle's citation hailed him as a "consumer of time and occupier of space, who more than anyone alse has demonstrated the need for science education." The person who delivered this speech appeared to be a nine-year-old girl.

Last year I promised you that I would return to Washington, and ask the President to join me in studying the calculus of Leibnitz and Newton, especially integrals, infinite series, eigenvectors, and the derivatives of simple continuous functions.

Today, I want to give you a brief history of the mathematical constant E. Euler's constant, better known as E, has the value 2 point 71828 dot dot dot. It is the limiting value of the sequence: parenthesis, one plus one over N, parenthesis, to the N, as N goes to infinity. It is also the value of X which satisfies the basic equation: the natural logarithm, L N of X, equals one.

Now that I know calculus, I have a new ambition I want to share with you. When I grow up, I want to be President of the United States. Thank you.

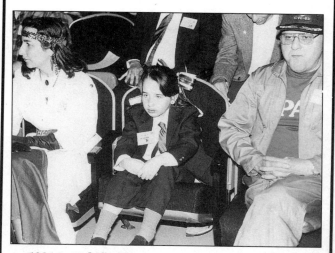

1991 Ig Nobel education Prize winner Dan Quayle (center) prepares to deliver the Returning Laureate's Address. Spam eater Dr. Jack S. Meagher is at right. A Quayle advisor is at left (photo: Roland Sharrillo).

INFORMATION FOR CONTRIBUTORS

The *Journal of Irreproducible Results* publishes original articles, news of particularly egregious scientific results, and short notices of satiric and humorous intent. The editors look forward to receiving your manuscripts, photographs, X-rays, drawings, etc. Please do not send biologic samples. Reports of research *results,* modest or otherwise, are preferred to speculative proposals. Submission of a manuscript implies that it is previously unpublished. Other material must be accompanied by an indication of source. All accepted manuscripts become the property of the Publisher. Manuscripts lost in transit will not be published.

The entire manuscript should be typed double-spaced on standard white bond paper, with generous margins all around, and submitted with a photocopy.

All illustrations should be submitted ready for press.

The Publisher does not redraw artwork. Line art is preferred to tone; whenever possible, please submit originals, not Xeroxes.

Because of high volume and low staffing, submissions cannot be acknowledged unless accompanied by a self-addressed, adequately stamped envelope.

To receive a list of arbitrary guidelines for authors, send a stamped, self-addressed envelope to: Guidelines for Contributors, % *JIR*, at the address below.

Please address your submissions to:

Marc Abrahams, Editor
The Journal of Irreproducible Results
% Wisdom Simulators, Inc.
P.O. Box 380853
Cambridge, MA 02238, USA

STAY AHEAD OF THE LATEST DEVELOPMENTS!

Subscribe to the *Journal of Irreproducible Results*

Please send me a year's subscription (6 issues)

Name _____

Address _____

City/State/Zip _____

Country _____

From (if gift)° _____
°A note indicating the source of the gift will be sent to the recipient.
Please call for group gift-subscription discounts

RATES

In the United States:
☐ Individuals $21.00
☐ Libraries $40.00

In Canada and Mexico:
☐ Individuals $27.50
☐ Libraries $46.00

All Other Countries:
☐ Individuals $43.00
☐ Libraries $62.00

Please enclose check or money order and return to:
Journal of Irreproducible Results, P.O. Box 380853, Cambridge, MA 02238 USA
(800) 759-6102   (617) 876-7000